CAMBRIDGE LIBRARY COLLECTION

Books of enduring scholarly value

Cambridge

The city of Cambridge received its royal charter in 1201, having already been home to Britons, Romans and Anglo-Saxons for many centuries. Cambridge University was founded soon afterwards and celebrates its octocentenary in 2009. This series explores the history and influence of Cambridge as a centre of science, learning, and discovery, its contributions to national and global politics and culture, and its inevitable controversies and scandals.

The Jurassic Rocks of the Neighbourhood of Cambridge

Cambridgeshire is rich in Jurassic fossils, with ichthyosaurs having been found at Mepal and Barrington since 2000, and ammonites and belemnites occurring plentifully near Grafham Water. In the later nineteenth century, Thomas Roberts, an assistant to the Professor of Geology at Cambridge, was involved in close study of the Jurassic rocks of the district. Despite the complexity arising from troublesome local geologic discontinuities, Roberts correlated the paleontological record in the Jurassic rocks of East Anglia with that of other British and European areas. It was this kind of careful and systematic work that provided the foundation for the modern understanding of the Jurassic period and the mass extinctions at the end of the Mesozoic era. Roberts's essay was originally published posthumously in 1892, and is now reissued for a new generation of readers and fossil-hunters.

Cambridge University Press has long been a pioneer in the reissuing of out-of-print titles from its own backlist, producing digital reprints of books that are still sought after by scholars and students but could not be reprinted economically using traditional technology. The Cambridge Library Collection extends this activity to a wider range of books which are still of importance to researchers and professionals, either for the source material they contain, or as landmarks in the history of their academic discipline.

Drawing from the world-renowned collections in the Cambridge University Library, and guided by the advice of experts in each subject area, Cambridge University Press is using state-of-the-art scanning machines in its own Printing House to capture the content of each book selected for inclusion. The files are processed to give a consistently clear, crisp image, and the books finished to the high quality standard for which the Press is recognised around the world. The latest print-on-demand technology ensures that the books will remain available indefinitely, and that orders for single or multiple copies can quickly be supplied.

The Cambridge Library Collection will bring back to life books of enduring scholarly value across a wide range of disciplines in the humanities and social sciences and in science and technology.

The Jurassic Rocks of the Neighbourhood of Cambridge

Being the Sedgwick Prize Essay for 1886

Thomas Roberts

CAMBRIDGE
UNIVERSITY PRESS

CAMBRIDGE UNIVERSITY PRESS

Cambridge New York Melbourne Madrid Cape Town Singapore São Paolo Delhi

Published in the United States of America by Cambridge University Press, New York

www.cambridge.org
Information on this title: www.cambridge.org/9781108002936

© in this compilation Cambridge University Press 2009

This edition first published 1892
This digitally printed version 2009

ISBN 978-1-108-00293-6

THE JURASSIC ROCKS OF CAMBRIDGE.

London: C. J. CLAY AND SONS,
CAMBRIDGE UNIVERSITY PRESS WAREHOUSE,
AVE MARIA LANE.

Cambridge: DEIGHTON, BELL AND CO.
Leipzig: F. A. BROCKHAUS.
New York: MACMILLAN AND CO.

THE JURASSIC ROCKS

OF THE

NEIGHBOURHOOD OF CAMBRIDGE.

BEING THE SEDGWICK PRIZE ESSAY FOR 1886.

BY THE LATE

THOMAS ROBERTS, M.A., F.G.S.,

ST JOHN'S COLLEGE, CAMBRIDGE.

CAMBRIDGE:

AT THE UNIVERSITY PRESS.

1892

Cambridge:
PRINTED BY C. J. CLAY, M.A. AND SONS,
AT THE UNIVERSITY PRESS.

PREFACE.

A FEW words of introduction are called for to explain the circumstances under which this essay was written and the reason why its publication has been so long delayed.

The liberal founder of the Sedgwick Prize left to the examiners considerable discretion in the choice of subjects, but it has been their general custom to propose some line of enquiry in which it might fairly be expected that the opportunities for original research and facilities for carrying it on were not out of reach of the student who was fresh from his University training and who was as yet unfettered by any obligations as to his line of study; so that the work proposed might contain in it the suggestion of a career as well as a test of power of original investigation.

The scheme has so far been eminently successful, and almost every winner of the Sedgwick Prize has become an authority upon the branch upon which he was invited to write.

Thus Mr Roberts, who, as Assistant to the Woodwardian Professor, was acquainted with the rich stores of material available for research in the Woodwardian Museum, had his attention more particularly turned to the correlation of the Jurassic Rocks of the Northern district of England with those of the South-West.

This enquiry was attended with considerable difficulty from the fact that throughout the greater part of the period the deposits were laid down under locally shifting geographical conditions, so that the district was from to time divided into different and changing hydrographical areas, the sediment varying as barriers disappeared or were introduced and the forms of life more or less readily yielding to the influence of external circumstances.

In this connection therefore the East Anglian district is especially interesting to geologists. In it that great change of direction in the lie of the Mesozoic rocks takes place which turns them from a North-Easterly to a Northerly strike; a change which is connected with axes of movement of far earlier date than Jurassic times.

Here there was an ancient pre-Jurassic land against which the Oolitic and Cretaceous rocks thin out. In this region as might be expected we find evidence that geographical changes affecting the conditions of life were still going on during the period of the deposition of the Jurassic Rocks.

And this is not all, for the upper members, which form such an important feature of the Oolitic series in the South-West, are missing in East Anglia; some have certainly been cut off by denudation so that the Cretaceous series creeps unconformably across the Kimeridge and Oxford and intermediate clays, and the highest beds seen under the Lower Greensand are never the top of the Oolitic series and indeed belong to very different parts of it in closely adjoining areas.

Here therefore all the skill of the palæontologist must be called in to the aid of the stratigraphical geologist to enable him to correlate the Northern with the Southern series, and trace palæontological continuity when the great lithological distinctions were no longer persistent.

This was the problem offered for solution in the Sedgwick Prize Essay, and this is the subject on which Mr Roberts has done so much good work since the Prize was awarded to him in 1886, and on which we had expected so much more from him had not his untimely end cut short the work. He himself always hoped to revise and extend the scope of the Essay, and in consequence its publication was too long delayed.

In this Essay Mr Roberts first gives a bibliographical and historical introduction and then treats of the various subdivisions of the Oolitic series in ascending order, beginning with the Oxford and ending with the Kimeridge Clay, the highest member of the series seen in this area.

At one time these two names included all the great Jurassic clays of the Fenland, and the Upware limestone was considered a lenticular development of the Coral Rag between them while the various limestone bands of Littleport and Knapwell were looked upon as subordinate beds in the upper division, and those of St Neots, St Ives, Elsworth, &c. were referred to the lower or Oxford Clay. As time went on, however, it was found possible to distinguish the intermediate portions of the Great Clay Series and correlate certain horizons by their fossils with zones already worked out elsewhere.

Mr Roberts, after much original work in the district and after having examined carefully the supposed equivalent formations in other areas, has summed the work already done and added much valuable matter as the result of his own researches. He concludes with two chapters on the correlation of the Jurassic Clays of East Anglia with those of other English areas and with those of the Continent.

This work therefore, with its full lists of fossils from each horizon and exact references to localities where sections were seen, is indispensable for the student of Cambridge geology and most valuable for all specialists in the Jurassic Rocks.

We fortunately secured an able editor in Henry Woods, B.A., F.G.S., Scholar of St John's College and Lecturer on Palæontology in the Woodwardian Museum, and we issue the work with the approval of one of the highest authorities on the subject treated of, W. H. Hudleston, M.A., F.R.S., President of the Geological Society, whose encouragement and assistance Mr Roberts always gratefully acknowledged.

T. McKENNY HUGHES.

WOODWARDIAN MUSEUM,
July 22, 1892.

CONTENTS.

LIST OF ILLUSTRATIONS.

LIST OF ILLUSTRATIONS

I. List of Works on the Jurassic Rocks of the Neighbourhood of Cambridge.

1836. Fitton, W. H.

Observations on some of the Strata between the Chalk and the Oxford Oolite in the South-east of England. *Trans. Geol. Soc.*, ser. 2, vol. IV., p. 103. [Bedfordshire and Cambridgeshire, pp. 269—308, 316—317.]

1846. Sedgwick, A.

On the Geology of the Neighbourhood of Cambridge, including the Formations between the Chalk Escarpment and the Great Bedford Level. *Rep. Brit. Assoc.* for 1845, *Sections*, p. 40.

1861. Porter, H.

The Geology of Peterborough and its Neighbourhood. Peterborough.

—— Sedgwick, A.

A Lecture on the Strata near Cambridge and the Fens of the Bedford Level. Privately printed.

Supplement [to 'A Lecture on the Strata,' etc.]. Privately printed.

—— Seeley, H. G.

Notice of the Elsworth Rock and of other New Rocks in the Oxford Clay, and of the Bluntisham Clay above them. *Geologist*, vol. IV., p. 460.

On the Fen-clay Formation. *Geologist*, vol. IV., p. 552, and *Ann. Mag. Nat. Hist.*, ser. 3, vol. VIII., p. 503.

R. E.   1

1862. SEELEY, H. G.

On the Elsworth Rock and the Clay above it. *Rep. Brit. Assoc.* for 1861, *Sections*, p. 132.

Notes on Cambridge Geology. 1, Preliminary Notice of the Elsworth Rock and associated Strata. *Ann. Mag. Nat. Hist.*, ser. 3, vol. x., p. 97.

1865. On the significance of the sequence of Rocks and Fossils : Theoretical Considerations on the Upper Secondary Rocks as seen in the Section at Ely. *Geol. Mag.*, vol. II., p. 262.

1867. FISHER, O.

On Roslyn or Roswell Hill Clay-pit, near Ely. *Proc. Camb. Phil. Soc.*, vol. II., p. 51, and *Geol. Mag.*, vol. v. (1868), pp. 407, 438.

—— SAUNDERS, J.

Notes on the Geology of South Bedfordshire. *Geol. Mag.*, vol. IV., pp. 154, 543.

1868. SEELEY, H. G.

On the Collocation of the Strata at Roswell Hole, near Ely. *Geol. Mag.*, vol. v., p. 347.

1872. BONNEY, T. G.

Notes on the Roslyn Hill Clay Pit. *Geol. Mag.*, vol. IX., p. 403.

On the section exposed at Roslyn Hill Pit, Ely. *Proc. Camb. Phil. Soc.*, vol. II., p. 268.

1875. Cambridgeshire Geology. Cambridge.

—— JUDD, J. W.

The Geology of Rutland and the parts of Lincoln, Leicester, Northampton, Huntingdon, and Cambridge, included in Sheet 64 of the one-inch map of the Geological Survey. *Mem. Geol. Survey.*

—— BLAKE, J. F.

On the Kimmeridge Clay of England. *Quart. Journ. Geol. Soc.*, vol. XXXI., p. 196. [Cambridgeshire, pp. 201, 211, 217—222.]

1877. BLAKE, J. F. AND W. H. HUDLESTON.

On the Corallian Rocks of England. *Quart. Journ. Geol. Soc.*, vol. XXXIII., p. 260. [Cambridgeshire, pp. 313—315.]

—— BONNEY, T. G.

The Coral Rag of Upware. *Geol. Mag.*, dec. 2, vol. IV., p. 476.

1877 SKERTCHLY, S. B. J.

The Geology of the Fenland. *Mem. Geol. Survey.*

1878. MILLER, S. H. AND S. B. J. SKERTCHLY.

The Fenland, Past and Present. Wisbech and London.

—— BLAKE, J. F. AND W. H. HUDLESTON.

The Coral Rag of Upware. *Geol. Mag.*, dec. 2, vol. v., p. 90.

1881. BLAKE, J. F.

On the Correlation of the Upper Jurassic Rocks of England with those of the Continent. Part I. The Paris Basin. *Quart. Journ. Geol. Soc.*, vol. XXXVII., p. 497. [Cambridgeshire, pp. 571, 580, pl. xxvi.]

—— PENNING, W. H. AND A. J. JUKES-BROWNE.

The Geology of the Neighbourhood of Cambridge. *Mem. Geol. Survey.*

1886. CARTER, J.

On the Decapod Crustaceans of the Oxford Clay. *Quart. Journ. Geol. Soc.*, vol. XLII., p. 542. [Contains a note on the Oxford Clay of St Ives, by T. Roberts, p. 543.]

1887. ROBERTS, T.

On the Correlation of the Upper Jurassic Rocks of the Swiss Jura with those of England. *Quart. Journ. Geol. Soc.*, vol. XLIII., p. 229. [Cambridgeshire, pp. 262, 264.]

1889. The Upper Jurassic Clays of Lincolnshire. *Quart. Journ. Geol. Soc.*, vol. XLV., p. 545. [Cambridgeshire, pp. 547, 555—559.]

1891. WHITAKER, W., H. B. WOODWARD, F. J. BENNETT, S. B. J. SKERTCHLY, AND A. J. JUKES-BROWNE.

The Geology of parts of Cambridgeshire and of Suffolk (Ely, Mildenhall, Thetford). *Mem. Geol. Survey.*

MAPS OF THE GEOLOGICAL SURVEY. 51, S.W. (Cambridge). 51, N.W. (Ely). 52, S.E. (St Neots). 52, N.E. (Huntingdon).

II. Introduction and History.

The town of Cambridge, and the district immediately adjoining, is underlain by rocks of Cretaceous age. To the west of this area, however, the Upper Jurassic rocks occupy a broad tract of low and uneven ground on both sides of the River Ouse from Bedford to Huntingdon: this continues northward, increasing in width, and dips under the Fen country of South Lincolnshire. South of Elsworth, these rocks also sweep round to the east and underlie the Fens of North Cambridgeshire.

The Jurassic rocks of this district have been described or referred to by various authors, all of whom agree in including the series in the Middle and Upper Oolites; considerable difference of opinion exists, however, as to their internal subdivision and arrangement.

In the south of England, where the Middle and Upper Oolites are typically developed, the main subdivisions are as follows:

Upper Oolites $\begin{cases} \text{Purbeck Beds.} \\ \text{Portland Sand and Stone.} \\ \text{Kimeridge Clay.} \end{cases}$

Middle Oolites $\begin{cases} \text{Corallian Beds.} \begin{cases} \text{Upper Calcareous Grit or Supracoral-} \\ \quad \text{line Beds.} \\ \text{Coral Rag.} \\ \text{Coralline Oolite.} \\ \text{Lower Calcareous Grit.} \end{cases} \\ \text{Oxford Clay, with the Kellaways Rock at its} \\ \quad \text{base.} \end{cases}$

The lowest member of the series, the Oxford Clay, is distributed in the form of a narrow band running across England in a north-easterly direction, from Dorsetshire to Yorkshire; it is fairly constant in its character throughout, though it exhibits a slight variation in the calcareous and arenaceous beds at its base, which are best developed in Yorkshire.

The Corallian Rocks of Dorsetshire are composed of limestones, clays and grits. Proceeding in a north-easterly direction they become much thinner and finally disappear, as a calcareous deposit, in Oxfordshire. In Bedfordshire, Cambridgeshire and Lincolnshire, these rocks are represented by a clayey deposit with some thin limestone bands. At Upware, in Cambridgeshire, however, there is a local development of the typical calcareous deposit of the south. In Yorkshire, again, the calcareous element predominates in the rocks of Corallian age. Hence, the two ends of the tract of country, along which these rocks are distributed, are similar, but great changes take place in the intermediate portion.

The Kimeridge Clay has much the same distribution as the Oxford Clay and is fairly persistent in its lithological character throughout.

The Portland and Purbeck beds are found only in the south and south-midland counties of England, with the possible exception of Yorkshire[1]; they are wholly wanting and probably never existed in the Cambridge district.

The Jurassic rocks represented in the neighbourhood of Cambridge are arranged as follows, in descending order:—

3. Kimeridge Clay.
2. Corallian Beds.
1. Oxford Clay.

The Corallian Rocks have, by some authors, been said to be absent from this district, with the exception of the local development at Upware. Professor Seeley, however, has shewn that the Corallian limestones of the south of England and Yorkshire are here represented by a clay. This Corallian clay, together with the underlying (Oxford) and overlying (Kimeridge) clays have been termed, by Professor Seeley[2], the "Fen-clay." The same author includes in the Oxford Clay three calcareous beds, which he arranges in the following order from above downwards:—

3. Elsworth Rock.
2. St Ives Rock.
1. St Neots Rock.

[1] J. W. Judd, *Quart. Journ. Geol. Soc.*, vol. XXIV. (1868), p. 218. G. W. Lamplugh, *ibid.*, vol. XLV. (1889), p. 575, and *Rep. Brit. Assoc.* 1890 (1891), p. 808.
[2] *Geologist*, vol. IV. (1861), p. 552.

The St Neots Rock, he considers to be low down in the Oxford Clay, and "not far removed from the zone of the Kelloway rock[1]"; but in the following year (1862) he states that from observations made in Bedfordshire, there appears to be a great thickness of clay beneath the lowest zone seen at St Neots[2]. The St Ives Rock is stated to be 130 feet below the Elsworth Rock[3]; and from palæontological evidence, Seeley regards the Elsworth Rock as the uppermost zone of the Oxford Clay, though he states that it lies directly underneath the argillaceous equivalent of the Coral Rag[4].

In Mr Lucas Barrett's geological map of Cambridgeshire, the Elsworth Rock is mapped as Upper Calcareous Grit, and the beds above as Kimeridge Clay; he was led to this view by the fact that *Ostrea deltoidea* occurs in this clay[5].

Seeley's arrangement of the limestones in the Oxford Clay was adopted by Professors Sedgwick and Bonney; but the latter author states[6] that the fauna of the St Ives Rock points to a rather high position in the Oxford Clay series.

Messrs Blake and Hudleston[7] consider that the evidence derived from the fossils of the St Ives Rock "seems to indicate that it belongs to some part of the age of the Lower Calcareous Grit or even higher"; and further, they take the Elsworth Rock to be an exceptional development of the Lower Calcareous Grit[8].

The St Ives Rock is mapped by the Geological Survey as Lower Calcareous Grit, and a rock which comes above that of Elsworth, is included in the same formation. In the Survey Memoir[9], the Lower Calcareous Grit is stated to occur also, between the Oxford and Kimeridge Clays at Papworth St Everard.

During the last few years large additions have been made to the collection of fossils in the Woodwardian Museum from the Elsworth and St Ives Rocks, and there seems sufficient evidence from palæontological considerations alone to prove :—

[1] *Geologist*, vol. IV. (1861), p. 553.
[2] *Ann. Mag. Nat. Hist.*, ser. 3, vol. x. (1862), p. 106.
[3] *Geologist*, vol. IV. (1861), p. 552.
[4] *Ann. Mag. Nat. Hist.*, ser. 3, vol. x. (1862), p. 108.
[5] A. Sedgwick, *Supplement*, (1861), p. 23.
[6] *Cambridgeshire Geology*, (1875), p. 11.
[7] *Quart. Journ. Geol. Soc.*, vol. XXXIII. (1877), p. 313.
[8] *Ibid.*, p. 384.
[9] W. H. Penning and A. J. Jukes-Browne, *Geol. Neighbourhood of Cambridge*, (1881), p. 6.

(1) That the Elsworth and St Ives Rocks are approximately of the same age.

(2) That they represent, in part at any rate, the Lower Calcareous Grit of other areas.

I would therefore propose to place both these rocks in the Lower Calcareous Grit and make the Oxford Clay terminate below them.

The clay formation, which represents the Corallian Rocks of other areas, has been described by Professor Seeley, and named by him the Bluntisham Clay[1], the Tetworth Clay[2], the Gamlingay Clay[3], and finally the Ampthill Clay[4]. I propose to retain the last name, although the locality after which it is named, is outside the district under consideration. Several bands of limestone occur in this clay, and Professor Seeley states[5] that they have been met with "in positions which render it probable that they indicate divisions corresponding to the Coral Rag and Calcareous Grit." These bands, however, are so thin and contain so few fossils, that I scarcely think such a division can be upheld.

The Ampthill Clay overlies the Elsworth Rock, but the line of division between it and the overlying Kimeridge Clay, has not as yet been drawn. I propose to place it at the base of a bed of phosphatic nodules which appears to be fairly constant in this district. The localities where it has been met with will be given later on. Much of the Ampthill Clay has been mapped by the Geological Survey as Oxford Clay.

The Corallian Rocks of Upware have been termed the "Upware Limestone" by Professor Seeley[6], and he places it above the clay equivalent to the Coral Rag. He gives the following classification:—

Kimeridge Clay.
Upware Limestone.
Clay equivalent to Coral Rag.
Oxford Clay.

[1] *Geologist*, vol. IV. (1861), p. 460.

[2] *Rep. Brit. Assoc.*, 1861, (1862), *Sect.*, p. 132, and *Ann. Mag. Nat. Hist.*, ser. 3, vol. VIII. (1861), p. 504, and *Geologist*, vol. IV. (1861), p. 553, and *Ann. Mag. Nat. Hist.*, ser. 3, vol. X. (1862), p. 107. A. Sedgwick, *Supplement*, (1861), p. 23.

[3] *Ann. Mag. Nat. Hist.*, ser. 3, vol. XX. (1867), p. 28.

[4] *Index to Aves, etc., in the Woodw. Mus.* (1869), p. 109.

[5] *Geologist*, vol. IV. (1861), p. 461.

[6] *Index to Aves, etc., in the Woodw. Mus.* (1869), p. xxvi.

Professor Bonney[1] includes the whole of the limestones at
Upware in the Coral Rag. Messrs Blake and Hudleston[2] make
out a twofold division of the beds thus:—

   2. Coral Rag.
   1. Coralline Oolite.

The Lower Kimeridge Clay is well developed in this district,
but I have been unable to detect any Upper Kimeridge, unless, as
stated by Professor Blake[3], the uppermost part of the section at
Roslyn Hill be such.

The Jurassic beds strike in a north and south direction in the
southern and western portions of the district, but east and west in
the northern. The overlying Cretaceous beds strike in a north-
east and south-west direction, and their lowest member, the
Lower Greensand, overlaps the different divisions of the Jurassic
series. Thus, near Ely we find it resting on the Kimeridge
Clay; at Great Gransden, Everton, and Gamlingay, on the
Ampthill Clay; and at Sandy, on the Oxford Clay.

The following is a tabular view of the subdivisions of the
Jurassic rocks which I propose to adopt for this district:—

   4. Kimeridge Clay { Upper. / Lower.

                         Upware section.
   3. Ampthill Clay . . . { Coral Rag. / Coralline Oolite.

   2. Lower Calcareous Grit. (The Elsworth and St Ives
      Rocks.)

   1. Oxford Clay.

[1] *Cambridgeshire Geology*, p. 19.
[2] *Quart. Journ. Geol. Soc.*, vol. xxxiii. (1877), p. 313 *et seq.*
[3] *Ibid.*, vol. xxxi. (1875), p. 201.

III. OXFORD CLAY.

THE tract of country occupied by the Oxford Clay in the neighbourhood of Cambridge, is much obscured by drift, and the only
places where the clay can be examined are in brick-pits, wells and
railway-cuttings.   These exposures are not very numerous, and on
this account the boundaries of the Oxford Clay are hard to define.
Its western limit is marked by the outcrop of the Lower Oolites of
East Northamptonshire, and runs in an irregular line from Bedford
northwards to Peterborough.   Its eastern boundary is somewhere
to the east of the course of the River Ouse from Bedford to Huntingdon : the clay-pits of Sandy, St Neots, and Godmanchester are
in this formation, but no exposures exist to the east of these localities.   It probably underlies the Elsworth Rock, near the village of
Elsworth, though it is not now seen there.   Leaving Godmanchester, the next exposure is in the brick-pits of St Ives, but I have
not been able to detect it at the surface further north or east.

The thickness of the Oxford Clay is pretty considerable, but
owing to the paucity of sections, it cannot be determined with any
degree of accuracy.   In the Survey Memoir[1] the thickness is estimated at 700 feet, but this is too great, since most of the Ampthill
Clay is included in the Oxford Clay.   Near St Ives Railway
Station, a well was dug in this clay to a depth of 150 feet, and
another at Bluntisham to 300 feet, but in neither case was the
clay pierced.

[1] W. H. Penning and A. J. Jukes-Browne, *Geol. Neighbourhood of Cambridge*,
(1881), p. 5.

Near Sandy, the Oxford Clay is overlain by the Lower Green-sand, and at St Ives it is succeeded by the Lower Calcareous Grit All the other sections seen shewed a capping of drift.

The Oxford Clay consists principally of bluish-grey, and some dark-blue tenacious clays. It contains several thin calcareous bands, which are usually called 'clunch' by the workmen. They are sandy argillaceous limestones, greyish or sometimes dark in colour; when weathered they are light grey. The bands are usually about a foot in thickness, but they may be less or a little more. At times they are nodular and discontinuous, forming a layer of more or less isolated calcareous nodules. They do not appear to be constant, and are, probably, only local in their development.

Crystals of selenite occur in the clays, but these are found principally in the upper part of the series. Nodules of iron pyrites are common, and fragments of fossil wood are occasionally found. Many of the fossils, and especially the Ammonites, are pyritized, but some have their shells composed of carbonate of lime. The interior of some of the fossils is occasionally filled with barytes. The clays are largely worked for brick-making, and it is in the brick-pits that the principal sections are seen.

The lowermost beds of the Oxford Clay do not occur in the Cambridge district, but they are met with in the neighbourhood of Peterborough and also in Northamptonshire. Professor Judd[1] gives the following localities in East Northamptonshire, where the Kellaways Beds are dug:—Oundle, Southwick, Benefield, Dogsthorpe, Uffington, Kate's Bridge, and Warmington. "These beds," he states[2], "which lie directly upon the Cornbrash, consist of an alternation of clays, usually light-coloured, very arenaceous, and sometimes pyritous, with irregular beds of whitish sand. The latter are not unfrequently cemented by calcareous matter into a friable rock, in which case they are usually full of fossils." The following are recorded from these beds:—*Gryphœa bilobata*, Sow., *Avicula inœquivalvis*, Sow., *Belemnites Oweni*, Pratt.

Professor Judd makes out the following succession in the Oxford Clay of that district:—

(*f*) Clays with Ammonites of the group *Cordati*.
(*e*)   „   „   „   *Ornati*.

---

[1] *Geology of Rutland*, (1875), p. 232.   [2] *Ibid.*, p. 232.

   (*d*)   Clays with *Belemnites hastatus.*
   (*c*)   ,,   *Belemnites Oweni.*
   (*b*)   ,,   *Nucula nuda.*
   (*a*)   Kellaways Sands, Sandstones, and Clays.

The lowermost zones of the Oxford Clay in the Cambridge district are exposed near St Neots, seven miles south-south-west of Godmanchester, where there are two brickyards. The first occurs at Eynesbury, just south of St Neots, and here there are two pits: the northern one, which may be called *Pit No. I.*, shews :—

		ft.	in.
(*a*)	Soil and flint gravel	2	0
(*b*)	Bluish-grey clay	2	6
(*c*)	Soft sandy limestone		7
(*d*)	Bluish-grey clay	4	6
(*e*)	Grey sandy limestone		6
(*f*)	Greyish clay becoming brown when weathered .	8	0
(*g*)	Sandy limestone		10
(*h*)	Dark-blue tenacious clay, which, at the time of my visit, was exposed to a depth of two feet: the workmen, however, informed me that this bed was 5 feet thick	5	0
	and that below it there was a thin bed of 'clunch,' beyond which they did not work. A trial boring was at one time made below this clunch, through 18 feet of dark-blue clay, when another bed of clunch was met with . .	18	0

The upper clays (*b*), (*d*) and (*f*) are much lighter in colour than the lower clays (*h*), but this I attribute chiefly to the weathering which the former have undergone. The beds dip to the south-west at a low angle. All the clays are fossiliferous, *Gryphœa dilatata* being more common in the lighter clays, whilst the Ammonites are found chiefly in the lower clays. The following were collected from this pit :—

          Ammonites athletus, Phil.
               ,,   *Duncani,* Sow.
               ,,   *Jason* (Rein.)
          Cucullœa concinna, Phil.
          Gryphœa dilatata, Sow.

The second pit (*Pit No. II.*) lies but a short distance south of the one described above ; in it was seen :—

		ft.	in.
(a)	Soil, gravel and drift	5	0
(b)	Bluish-grey clay	2	6
(c)	Sandy limestone		7
(d)	Bluish-grey clay	3	0

The limestone (c) in this pit is the same as (c) in Pit No. I.

The second brickyard is about a mile further south, and is situated on the east side of the road leading to Little Barford. The following section was seen in this pit (*Pit No. III.*):—

		ft.	in.
(a)	Soil and flint gravel	2	0
(b)	Greyish-brown clay	7	0
(c)	Greyish sandy limestone, much jointed and somewhat nodular; in some portions of the section it is distinctly divided into two beds, but this division is not constant. It contains some decomposed pyrites, and the fossils are not abundant	11–12	
(d)	Blue tenacious clay	17	0
(e)	Soft sandy limestone, containing a quantity of oxide of iron	7–9	
(f)	Blue clay—only two feet of this was visible, but I was informed that it extended down to a depth of 4 feet 6 inches . . .	4	6

and that below it comes :—

    (1) A thin bed of 'clunch,' somewhat harder than the bed (e).

(2)	Blue clay	8	0

    (3) Bed of 'clunch,' which was not pierced.

The following fossils were obtained :—

    (i) From the limestone (c).

> *Ammonites* (fragment only).
> *Cerithium muricatum* (Sow.)
> *Astarte robusta*, Lyc.
>   „   sp.
> *Gryphæa dilatata*, Sow.

(ii) From the limestone (*e*).

> *Ammonites Duncani*, Sow.
> *Astarte robusta*, Lyc.
> *Cucullœa concinna*, Phil.
> *Rhynchonella varians* (Schloth.)

(iii) From the clays.

> *Pliosaurus Evansi*, Seeley.
> *Ammonites athletus*, Phil.
> „   *Jason* (Rein.)
> *Gryphœa dilatata*, Sow.
> *Rhynchonella varians* (Schloth.)

The following fossils, labelled St Neots, in the Woodwardian Museum, come from one or other of the above brickyards:—

> *Belemnites hastatus* (Montf.)
> „   *obeliscus*, Phil.
> „   *Puzosianus*, d'Orb.
> „   „   var. *verrucosus*, Phil.
> *Ammonites Duncani*, Sow.
> „   *Jason* (Rein.)
> „   sp.
> *Trigonia costata*, Sow.
> *Pecten fibrosus*, Sow.

The uppermost bed of limestone (*c*) in Pit No. III. is the one which has been called the St Neots Rock by Professor Seeley. He states[1] that " near Eynesbury, in laying the plates of the Great Northern Railway, a rock was cut down to.   The uppermost band of about 8 inches was removed: it held beneath a large quantity of water, which produced the singular effect of giving all the workmen who drank of it the ague. Below this bed of stone was another of clay, and then another of the rock; but whether there were any further alternations was not determined. The rock is described as having the aspect of Cornbrash." The clay-pit (No. III.) cuts into the ridge, through which the railway-cutting mentioned by Professor Seeley passes, and is worked to within a yard of the railway boundary, so that the strata which were then exposed in the cutting, but are now obscured by vegetation, may

---

[1] *Ann. Mag. Nat. Hist.*, ser. 3, vol. x. (1862), p. 105.

be well seen in the clay-pit adjoining. The limestone band (c) is at such a depth that it would be reached at the bottom of the cutting, so that it is doubtless the St Neots Rock of Professor Seeley. From the description given above, and also from the fossils, it is seen that this rock does not differ markedly from the other calcareous bands which occur throughout the Oxford Clay; on this account I hardly think it necessary to retain the term "St Neots Rock" for it.

By comparing the sections given above in Pit No. I. and Pit No. III., it seems highly probable that the beds c, d, e, f, g of the former are the same as e, f, f_1 and f_2 of the latter, the thickness of the beds in the one pit being precisely the same as those in the other. Summing up the results obtained from the three pits, the following would appear to represent the whole development of the Oxford Clay in the St Neots district :—

		ft.	in.
(1)	Greyish-brown clay	7	0
(2)	Greyish sandy limestone ("St Neots Rock")		11–12
(3)	Bluish clay	17	0
(4)	Sandy limestone		7–9
(5)	Blue clay	4	6
(6)	Sandy limestone		6
(7)	Blue clay	8	0
(8)	Sandy limestone		10
(9)	Blue clay	5	0
(10)	Sandy limestone		6 (?)
(11)	Blue Clay	18	0
(12)	Sandy limestone		not pierced.

The *Cordati* group of Ammonites, so common in the uppermost zones of the Oxford Clay, are absent at St Neots. Its characteristic fossils are the *Ornati* Ammonites, and the St Neots clays should be placed in zone *e* of Professor Judd.

Further south, the Oxford Clay is worked in a pit north of Sandy station. The section shows about 7 feet of blue clay, and near its base is a bed of grey sandy limestone from 8 to 9 inches thick; about 5 feet above the latter is an irregular layer of calcareous nodules, some of which show a septarian structure. The clay contains iron pyrites and crystals of selenite, and yielded the following fossils :—

Belemnites hastatus (Montf.).
       „   *Puzosianus*, d'Orb.
Ammonites cordatus, Sow.
       „   *Mariæ*, d'Orb.
       „   *trifidus*, Sow.
Alaria trifida (Phil.)
Nucula ornata, ? Quenst.
Avicula inæquivalvis, Sow.
Gryphæa dilatata, Sow.

The siding joining this pit with the main line cuts through a slight ridge, where the Oxford Clay, with two limestone-bands, is seen to be unconformably overlain by the Lower Greensand.

At Godmanchester, about three miles west of St Ives, is a brick-yard, which is probably at a somewhat lower horizon in the Oxford Clay than that of St Ives. The section is as follows :—

(a)   Boulder clay, which is largely made up of the
       Jurassic clays, but contains septarian nodules
       and pebbles of flint and chalk. It is called
       'callers' by the workmen   .   .   .   . 18 feet.

(b)   Dark-blue tenacious clays, which contain no
       selenite crystals, but small cylindrical masses
       of iron pyrites are occasionally found. Near
       the top of the clay is a thin bed of 'clunch';
       it is a greyish sandy limestone, about 4
       inches thick: its depth below the surface of
       the clay varies from 3 to 6 feet   .   . 24 feet seen.

The following fossils were obtained from the clay :—

Belemnites Puzosianus, d'Orb.
Ammonites Achilles, d'Orb.
       „   *Bakeriæ*, Sow.
       „   *cordatus*, Sow.
       „   *oculatus*, Phil.
Gryphæa dilatata, Sow.

The most easily accessible section from Cambridge is in the large brickyard which lies to the west of the town of St Ives, where the following beds (fig. 1) may be seen :—

FIG. 1. *Section in Clay-pit west of St Ives*[1].

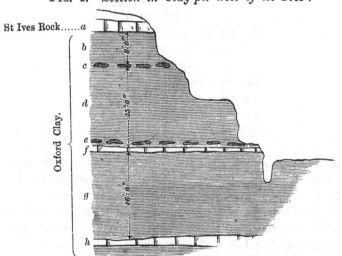

(a) A brown ferruginous limestone—the St Ives Rock. The upper part of the clay immediately underlying it has already been worked, and the rock is only seen a little to the west of the pit, where the clay is now being dug 3 0
(b) Dark-blue clay 8 0
(c) Layer of calcareous nodules . . . 0 9
(d) Dark-blue clay with selenite crystals . . 15 0
(e) Layer of calcareous nodules, discontinuous in the eastern portion of the pit, but more regular in its western part . . . 9–10
This is separated by a thin layer of clay from
(f) A bed of greyish argillaceous limestone. It is not very constant in character, since it becomes more irregular in the western part of the pit 1 0
(g) Blue clay which has been worked to a depth of 16 feet, where, the workmen informed me, a calcareous bed, a few inches in thickness, was met with 16 0

[1] For the use of this illustration we are indebted to the Council of the Geological Society. It was originally published in a note by the author contained in a paper by Mr James Carter 'On the Decapod Crustaceans of the Oxford Clay,' *Quart. Journ. Geol. Soc.*, vol. XLII. (1886), p. 542.

Fossils are fairly abundant in the clay, the most common being *Gryphœa dilatata*, which occurs throughout the clay and the calcareous bands. *Waldheimia impressa* is most abundant immediately above the lowermost limestone (*h*). The subjoined list of fossils from this pit is compiled from the specimens in the Woodwardian Museum.

LIST OF FOSSILS FROM THE OXFORD CLAY OF ST IVES.

CEPHALOPODA.

Belemnites abbreviatus, Mill.
„ *hastatus* (Montf.)
„ *Puzosianus*, d'Orb.
Belemnoteuthis, sp.
Nautilus calloviensis, Oppel
Ammonites Achilles, d'Orb.
„ *athletus*, Phil.
„ *Babeanus*, d'Orb.
„ *Bakeriœ*, Sow.
„ *cordatus*, Sow.
„ *crenatus*, Brug.
„ *Eugenii*, Rasp.

Ammonites excavatus, Sow.
„ *flexuosus*, Münst.
„ *Goliathus*, d'Orb.
„ *Hecticus* (Rein.)
„ *Jason* (Rein.)
„ *lophotus*, Ziet.
„ *Mariœ*, d'Orb.
„ *oculatus*, Phil.
„ *perarmatus*, Sow.
„ *rupellensis*, d'Orb.
„ *trifidus*, Sow.
„ n. sp.

GASTEROPODA.

Cerithium Damonis, Lyc.
„ sp.

Alaria trifida (Phil.)

LAMELLIBRANCHIATA.

Pholadomya Phillipsi, Morr.
Thracia depressa (Sow.)
Astarte, sp.
Isocardia, sp.
Cardium Crawfordi, Leck.
Trigonia clavellata, Sow.
„ *elongata*, Sow.
„ *Pellati*, Mun. Chal.
Nucula elliptica, Phil.
„ *nuda*, Phil.
„ *ornata*, Quenst.
„ *turgida*, Bean

Leda lacryma (Quenst.)
„ sp.
Cucullœa concinna, Phil.
Modiola bipartita, Sow.
Pinna mitis, Phil.
Perna, sp.
Avicula inœquivalvis, Sow.
Lima rigida (Sow.)
„ sp.
Ostrea gregaria, Sow.
Exogyra nana (Sow.)
Gryphœa dilatata, Sow.

BRACHIOPODA.

Rhynchonella lævirostris, Mc Coy   Waldheimia impressa (Von Buch)
    „   varians (Schloth.)   Terebratula oxoniensis [Walker MS.]
                                 Dav.

VERMES.

Serpula, sp.   Vermilia sulcata, Sow.

CRUSTACEA.

Eryon sublevis, Carter   Glyphea Regleyana, Meyer
Eryma Mandelslohi (Meyer)   Magila Pichleri, Oppel
   „   ventrosa (Meyer)   „   levimana, Carter
   „   Villersi, Morière   „   dissimilis, Carter
   „   Babeaui, Etallon   Mecochirus socialis (Meyer)
   „   Georgei, Carter   Goniochirus cristatus, Carter
   „   ?pulchella, Carter   Pseudastacus, sp.
Glyphea hispida, Carter   Pagurus, sp.

ECHINODERMATA.

Acrosalenia, sp.   Pentacrinus, sp.

The sections described above are the only ones which, so far as I know, occur in the district under consideration. In them three well-marked palæontological zones can be distinguished :—

3. Zone of *Ammonites Duncani* and *Jason* (the *Ornati* group of Ammonites) of St Neots.

2. Zone of *Waldheimia impressa*, at the base of the St Ives clay-pit.

1. Zone of *Ammonites perarmatus* (rare), *A. crenatus* and *A. oculatus*, and the *Cordati* group of Ammonites, of the St Ives clay-pit.

IV. Lower Calcareous Grit.

1. *The Elsworth Rock.*

THE village of Elsworth is four and a half miles south of St Ives, and about eight miles west-north-west of Cambridge. A stream runs through the village in a north and south direction, which has excavated a shallow valley in the Oolite clays. Along the banks of this stream, more especially to the south of the road which runs through Elsworth towards Papworth St Agnes, a limestone is exposed to which the name Elsworth Rock has been given. It is best seen in the road which runs along the east side of the stream. Near the southern end of the village it crops out in the bed of the stream itself; it also forms the floor of a pond in a field immediately to the south of the village. At "t" in "Elsworth Rock" on the Survey Map (sheet 51 S.W.) about a mile north of Elsworth, a ferruginous oolitic limestone is seen in the sides of a pond, this limestone may be the Elsworth Rock. These are the only places where it is now visible. Where exposed in the roadside the limestone is of a reddish-brown colour, and is highly charged with ferruginous oolitic grains; it is very hard and much jointed at the surface, but when dug into it is more flaggy. Professor Seeley had, at one time, a pit sunk into the limestone near its outcrop, and the following account is taken mainly from his description. A clay was cut through to a depth of $6\frac{1}{2}$ feet. Then the rock was reached, which is described[1] as "a dark-blue homogeneous limestone, which I can compare to nothing but the unseptarious cement-stones of the clays. The oolitic grains were abundant and as deeply ferruginous as though they had been exposed to the air; while scattered irregularly about, branching and interlacing, were masses of undecomposed iron pyrites." The rock was

[1] *Ann. Mag. Nat. Hist.*, ser. 3, vol. x. (1862), p. 99.

exceedingly hard, and much difficulty was experienced in digging into it. The thickness of the rock is given as 3 to 4 feet, and in some places 7 feet. This variable thickness is said to be due to denudation prior to the deposition of the clay which overlies it. This clay is of a brown-black colour, is 5 feet thick, and contains numerous specimens of a small variety of *Ostrea Marshi.* Above the clay comes another rock 1 foot 6 inches thick, which is stated to have "all the outer characters of that below, being equally rusty and quite oolitic—except at one point, where, getting a section in a ditch, it was found to be yellowish white, more sandy, and almost free from oolitic particles. The fossils here were few, but, under the slightly altered conditions, differed a little from those met with in other places. In general, they are very similar to those of the inferior deposit, though including some forms, which, as far as yet explored, have served to distinguish it[1]." Professor Seeley[2] thus makes a three-fold division of the Elsworth Rock series :—

							ft.	in.
(1)	An upper rock	.	.	.	.	.	1	6
(2)	A middle clay	.	.	.	.	.	5	0
(3)	A lower rock	.	.	.	.	.	**3–7**	

He gives no fossil lists from the upper rock, but he states[3] that there are numerous masses of *Serpulæ* in it, and that it differs from the lower beds "in the Brachyurous Crustacea, in the species of *Cidaris,* in the enormous size of *Gryphæa dilatata* (some being 11 inches long), the absence of *Gryphæa elongata,* the presence of plates of a Star-fish, the greater abundance of *Littorina perornata* and of *Pleurotomaria reticulata,* species of *Perna,* many *Ostrea gregaria.*" A list of fossils from the lower rock will be given later on.

A well was recently dug near the school at Elsworth, which is situated just above the outcrop of the Elsworth Rock, and in it two limestone beds were pierced: these were probably the upper and lower rocks referred to above.

A little more than three miles south-south-west of Elsworth

---

[1] Seeley, *op. cit.,* pp. 99, 100.

[2] *Geologist,* vol. IV. (1861), p. 460.

[3] *Ann. Mag. Nat. Hist.,* ser. 3, vol. x. (1862), p. 109.

(near Bourn) a well was dug, in which the following beds[1] were pierced :—

 (*a*) Dark-blue clay from which many Ammonites
    were obtained . . . . . . 84 feet

 (*b*) Alternating bands of stone and sand and an
    extremely hard grey-blue rock which proved
    very difficult to pierce . . . . 14 feet

The latter was probably a continuation of the Elsworth Rock, and Professor Seeley considers[2] that the sand replaces the middle clay of the Elsworth Rock series.

In the bottom of the railway cutting, west of Bluntisham, and about two miles north-north-east of St Ives, a rock was met with which was so hard that it had to be blasted in laying the drain on each side of the railway. Since then the rails have been raised a couple of feet, so that the rock is not now visible. I had a pit sunk to a depth of five feet, near the bridge which spans this cutting, without meeting the rock. Professor Seeley, however, was fortunate in securing a piece of it, and he states[3] that it contains "iron-shot oolitic grains and shells, quite resembling the rock of Elsworth." He gives no list of the fossils occurring in this rock. The clay which is now seen in the cutting is certainly of Ampthill Clay age, and it is probable that the rock may be on the same horizon as that of Elsworth.

2. *The St Ives Rock.*

As already stated, this rock forms the summit of the clay-pit in the St Ives brickyard. A small exposure of it may now be seen in the western part of the pit, some distance above the clays that are being worked. The rock is overlain by a thin capping of greyish-blue clay, in which I have been unable to find any fossils. The rock itself is of a brownish colour, and is made up of the following beds :—

---

[1] H. G. Seeley, *Ann. Mag. Nat. Hist.*, ser. 3, vol. x. (1862), p. 100.

[2] *Geologist*, vol. IV. (1861), p. 461.

[3] *Ann. Mag. Nat. Hist.*, ser. 3, vol. x. (1862), p. 101.

ft. in.

(a) Yellowish-brown calcareous clay, thin and irregularly bedded. Fossils not abundant 7

(b) Two thin beds of brown ferruginous limestone, the uppermost containing decomposed nodules of iron pyrites. Fossils, *Exogyra nana*, *Goniomya literata*, etc. 1 6

(c) Brown sandy fossiliferous limestone, containing oolitic grains of oxide of iron, like those in the Elsworth Rock. It is very nodular towards the base, and here the fossils are most abundant . 6

(d) Grey calcareous sandy clay, brown in its upper part, but below it passes down into the Oxford Clay. It contains *Pinna*, *Myacites recurva*, *Gryphœa dilatata*, *Pleurotomaria*, etc. . . 5

I have, so far, been unable to see the rock in its unweathered condition, though I have worked and quarried a considerable quantity of it. It appears to dip to the east at a low angle.

The resemblance of the lower part of the St Ives Rock to that of Elsworth is very striking, and from the list which follows (p. 25), it will be seen that a large percentage of the fossils are common.

To the north of St Ives and a little to the east of the brickyard above referred to is another pit, where "a rock had formerly existed, but it was now all removed at the surface, having dipped down into the clay to the east of the pit. And still further to the east, at the point where the roads branch to Needingworth and Somersham, it appeared that, in another brickyard, they sometimes, when digging, came down upon a floor of hard stone, which they have never attempted to get through. To one going over the ground, the conclusion is irresistible that the rocks mentioned at all these pits [*i.e.* the St Ives pit and the two to the east of it] are one deposit dipping down to the east[1]."

Mr James Carter, of Cambridge, had in his collection some fossils, which he has lately presented to the Woodwardian Museum, and which he believes came from Holywell, a mile and a half east of St Ives. These fossils are precisely similar to those of the St Ives Rock, and their matrix is also identical. If they were collected at Holywell it would prove that a rock in every way similar to

---

[1] H. G. Seeley, *Ann. Mag. Nat. Hist.*, ser. 3, vol. x. (1862), p. 101.

that of St Ives, extends to that village and had been worked there. Professor Seeley has also examined this collection of fossils, and he considers[1] that the fossils and the lithological character of the matrix in which they occur, are like those of Elsworth. A list of those fossils will be given later on, and it will be seen that their affinity is much nearer to those of the St Ives Rock, than to those of Elsworth.

At Conington, three and a half miles south of the St Ives clay pit, a well was bored to a depth of 250 feet. A rock 5 feet thick was pierced at 100 feet below the surface. This rock Professor Seeley considers to be the same as the St Ives Rock. No description of the rock is given, neither are there any fossils recorded from it. The dip of the rocks (*i.e.* those of Elsworth) which are said to come above it furnishes the only evidence in favour of Seeley's view[2].

In the northern part of the village of Swavesey, a well was dug to a depth of 23 feet, when a hard brownish limestone was met with, and which was excavated to a depth of 2 feet (one of the workmen said 3 feet) but not pierced. This is probably a continuation of the St Ives Rock, as no band of this character and thickness occurs in the Oxford or Ampthill Clays.

During the early part of this year (1885) a well-boring was made at Chettering Farm, half a mile south-east of Stretham Ferry, and two and a half miles north-west of Upware. The beds which were pierced are given in the following list :—

		ft.	in.	ft.	in.
(a)	Peat	5	0		
(b)	Sand	3	6		
(c)	Lower Greensand	13	6		
				22	0
(d)	Clays with subordinate limestone bands; the latter were about 1 ft. thick and intervening clays 3 ft. 6 in. The clay contained some phosphatic nodules . .	36	0		
(e)	Laminated clays in which *Ammonites serratus* was found	12	0		
(f)	Clays with black phosphatic nodules . .	3	0		
				51	0

[1] *Ann. Mag. Nat. Hist.*, ser. 3, vol. x. (1862), p. 110.   [2] *Ibid.*

		ft.	in.	ft.	in.
	Brought forward			73	0
(g)	Clay with iron pyrites: near its base a bed of dark sandy limestone 2 ft. thick was met with	26	0		
				26	0
(h)	"Hard stone"—a greyish limestone crowded with oolitic grains of oxide of iron .	8	0		
(i)	Light-brown sandstone	11	0		
				19	0
(k)	Alternating bands of clay and limestone the latter 8 to 9 in. thick	12	0		
(l)	Clays with some fragments of fossils .	57	0		
(m)	Argillaceous limestone	2	0		
(n)	Clay with small pyritized Ammonite and *Ostrea*, sp.	10	0		
				81	0
	Total			199	0

This boring was visited just before it was abandoned, and the cores were examined: they were only two inches, and those from near the bottom only one inch, in diameter, consequently very few fossils could be obtained. The lowermost eighty-one feet, I consider to be Oxford Clay, although the only fossils which could be identified came from near the base.

The limestone and sandstone, (h) and (i), were broken through by means of a drill, and only small fragments of them were brought to the surface. I secured a specimen of *Pholadomya æqualis*, which probably came from the bed (h) and its matrix shows the oolitic grains and other characters precisely similar to the Elsworth and St Ives Rocks. It differs from them, however, in being thicker, and more especially in the sandy layer coming on below the limestone. It has been already stated that the rock met with in the well at Bourn, which there is reason to believe is a continuation of the Elsworth Rock, differs from it both in thickness and to some extent in lithological character. A similar difference exists between these limestones at Elsworth and St Ives. There is no doubt therefore but that the rock in the Chettering boring belongs to the Elsworth Rock series.

From these considerations it is seen that the Lower Calcareous Grit of this neighbourhood is somewhat variable in character. But, wherever it has been met with, it always contains a quantity of ferruginous oolitic grains disseminated through the limestone.

The accompanying table contains a list of the fossils from the Elsworth and St Ives Rocks, and also those which Mr Carter believes to have come from Holywell. In the first column I have indicated those which pass down into the Oxford Clay and older beds; in the fifth those which occur in the Lower Calcareous Grit of other areas; and in the sixth column those which are found in still higher beds in England.

LIST OF FOSSILS FROM THE ST IVES AND ELSWORTH ROCKS.

	Oxf. Clay and older beds	Elsworth Rock	St Ives Rock	Holywell	Low. Calc. Grit	Newer beds	REMARKS.
PISCES.							
Hybodus, sp.		×					Fragment only.
„ *grossiconus*, Ag.	×		×		×		
CEPHALOPODA.							
Belemnites Oweni ?, Pratt		×					Fragments only—referred to *B. tornatilis* by Seeley.
„ *hastatus* (Montf.)	×	×	×		×		
Nautilus perinflatus, Foord and Crick	×	×					
Ammonites Achilles, d'Orb.		×			×		Characteristic of *Étage Corallien* of France.
„ *canaliculatus*, Münst.		×			×		*Étage Oxfordien* of France.
„ *convolutus*, Quenst.	×	×			×		
„ *cordatus*, Sow.	×	×	×	×	×	×	
„ *Goliathus*, d'Orb.		×				×	*Étage Oxf.*, France. Coralline Oolite of England.
„ *Henrici*, d'Orb		×	×	×			*Étage Oxfordien* of France.
„ *perarmatus*, Sow.	×	×	×		×		
„ *planicordatus*, Seeley, MS.		×	×	×			
„ *plicatilis*, Sow.		×			×	×	
„ *vertebralis*, Sow.	×	×	×	×	×	×	
GASTEROPODA.							
Natica Clymenia, d'Orb.		×				×	Coralline Oolite of Yorkshire, *Oxf. infr.* France.
„ *Calypso*, d'Orb., var. *tenuis*, Hudl.		×	×	×	×		
Nerinea, sp.		×					
„ sp.			×				
Alaria bispinosa (Phil.)	×				×	×	
Turbo Meriani, Goldf.		×	×	×			
Pleurotomaria granulata, Lyc. *non* Sow.	×	×	×	×	×		

	Oxf. Clay and older beds	Elsworth Rock	St Ives Rock	Holywell	Low. Calc. Grit	Newer beds	REMARKS
Pleurotomaria Münsteri, Rœmer		✕	✕	✕	✕	✕	
,, sp.		✕					
Patella cf. mosensis, Buv.		✕			✕		
LAMELLIBRANCHIATA.							
Goniomya literata (Sow.)	✕	✕	✕	✕	✕	✕	
Myacites recurva (Phil.)	✕	✕	✕	✕	✕	✕	
,, Jurassi, Brong.	✕	✕	✕	✕	✕	✕	
Pholadomya parcicosta, Ag.		✕			✕	✕	*Corallien supr.*, France.
,, æqualis, Sow.		✕	✕	✕	✕	✕	
,, cingulata?, Ag.			✕				
,, concentrica, Rœm.		✕			✕		Upper Coral Rag of Spitzhuts, near Hildesheim.
Thracia depressa (Sow.)	✕	✕	✕	✕	✕	✕	
Anatina undulata (Sow.)	✕		✕		✕	✕	
Astarte ovata (Smith)		✕		?		✕	
,, robusta, Lyc.	✕	✕	✕			✕	Kellaways Rock and Cornbrash.
,, sp.		✕					
Isocardia globosa, Bean		✕	✕	✕			
Opis angulosa, d'Orb.		✕				✕	
Myoconcha cf. Sœmanni, Dollf.		✕	✕	✕		✕	
Lucina globosa, Buv.	✕	✕				✕	
,, ampliata?, Phil.			✕	✕		✕	
Unicardium depressum (Phil.)	✕	✕				✕	
Cardium Crawfordi, Leck.	✕	✕	✕	✕			Kellaways Rock.
Trigonia elongata, Sow.	✕	✕	✕			✕	
,, perlata, Ag.		✕				✕	
,, Hudlestoni, Lyc.		✕				✕	
Nucula, sp.		✕	✕				
Arca æmula, Phil.		✕			✕	✕	
,, terebrans, Buv.		✕				✕	
,, sp. (with coarse ribs)		✕					
,, sp.			✕				} Casts only found.
,, sp.			✕				
Cucullæa clathrata, Leck.	✕	✕					Lower Oolites.
,, oblonga, Sow.	✕	✕			.	✕	
,, elongata, Phil.		✕				✕	
Lithodomus, sp.		✕	✕				Boring in *Gryphæa dilatata*.
Modiola bipartita, Sow.	✕	✕	✕	✕	✕	✕	
,, cancellata, Rœm.			✕			✕	
Pinna lanceolata, Sow.	✕	✕	✕		✕	✕	
,, mitis?, Phil.		✕					Fragment only.
,, sp.		✕					
Perna mytiloides, Lam.		?	✕		✕	✕	
Avicula pteropernoides, Blake and Hudl.		✕				✕	*Avicula pterosphena*, Seeley, MS.
,, ovalis, Phil.	?	✕			✕	✕	
,, expansa, Phil.	✕	✕	✕	✕	✕	✕	

	Oxf. Clay and older beds	Elsworth Rock	St Ives Rock	Holywell	Low. Calc. Grit	Newer beds	REMARKS.
Avicula inæquivalvis, Sow.	x	x				x	
„ braamburiensis, Sow.	x	x			x		
Lima pectiniformis(Schloth.)	x	x				x	
„ duplicata, Sow.	x	x				x	
„ læviuscula, Sow.		x	x		x	x	
„ rigida (Sow.)	x	x	x			x	
„ n. sp.		x					
Pecten lens, Sow.	x	x	x	x	x	x	
„ vagans, Sow.	x	x	x	x		x	
„ articulatus (Schloth.)		x	x	x		x	
Hinnites abjectus (Phil.)	x	x					
„ Sedgwicki, Seeley MS.		x					
„ velatus (Goldf.)			x			x	
Plicatula fistulosa, Lyc.	x	x				x	
Anomia, sp.			x				
Placunopsis, sp.		x					
Ostrea gregaria, Sow.		x	x		x	x	
Gryphæa dilatata, Sow.	x	x	x		x	x	
Ostrea flabelloides, Lam.	x	x	x		x	x	
Exogyra nana (Sow.)	x	x	x		x	x	
Ostrea discoidea, Seeley		x				x	
BRACHIOPODA.							
Terebratula insignis, Schüb.		x				x	
Waldheimia bucculenta (Sow.)		x	x		x		
„ Hudlestoni (Walk.)			x		x		
VERMES.							
Serpula, sp.		x	x	x			
CRUSTACEA.							
Glyphea, sp.		x					
Goniocheirus cristatus, Carter			x				
ECHINODERMATA.							
Apiocrinus, sp.		x					
Pentacrinus, sp.		x					
Millericrinus echinatus (Schloth.)		x			x		
Cidaris Smithi, Wright		x					
„ florigemma, Phil.			x		x	x	
Pseudodiadema versipora (Woodw.)		x	x		x	x	
Holectypus depressus (Leske)	x	x	x	x	x	x	
Collyrites bicordata (Leske)		x	x	x	x		
ACTINOZOA.							
Thecosmilia, sp.		x					

3. *Palæontological affinities of the fauna of the St Ives
and Elsworth Rocks.*

(i) *To each other :—*

The number of species given from these rocks is as follows :—

                        From Elsworth, 85   From St Ives, 54
Deduct from both the doubtful species and
    those of which only the generic name
    is given 20   9
                        ⎯⎯⎯   ⎯⎯⎯
                         65   45

Of these 45 undoubted species found at St Ives, 35 occur in
the Elsworth Rock, that is to say 77 per cent. are common to
both. It has already been shown that the two rocks differ somewhat in their lithological character and thickness; and it is very
probable that the physical conditions under which they were
deposited were somewhat different, since there is more calcareous
material in the Elsworth Rock than in that of St Ives, and the
latter contains more argillaceous and arenaceous admixture than
the former; this might account for the slight difference in the
fauna of the two localities. The most obvious distinction between
the two, is in the greater number of echinoids at St Ives; of the
four species which occur there, only one (*Pseudodiadema versipora*)
has been found at Elsworth, and this is represented by a single
specimen; at St Ives, on the other hand, *Holectypus depressus*
and *Collyrites bicordata* are fairly common. Taking all these
facts into consideration, I maintain that the palæontological
evidence is strongly in favour of both rocks being on the same
geological horizon. Unfortunately there is no exposure in the
clays immediately below the rock at Elsworth, so that no evidence
can be obtained from the underlying beds; the same may be said
of the overlying beds.

The Holywell fossils are practically identical with those of the
St Ives Rock: only one species occurs in the list from the first-
named locality which has not been found at St Ives.

(ii) *To the Oxford Clay and other beds :—*

The following five species unite the Elsworth and St Ives

Rocks, with the Oxford Clay and older beds; none of them are found in beds higher than the Oxford Clay in England:—

Astarte robusta, Lyc. (Cornbrash, Kellaways Rock and Oxford Clay).

Isocardia globosa, Bean (Cornbrash).

Cucullœa clathrata, Leck. (Great Oolite).

Cardium Crawfordi, Leck. (Kellaways Rock).

Hinnites abjectus, (Phil.) (Great Oolite).

The majority of these range down into the Lower Oolites, but the list contains no fossil which is characteristic of and peculiar to the Oxford Clay proper.

Two species of Ammonites (*A. Henrici* and *A. canaliculatus*) are given in the above table which are recorded from the Oxfordian of France. It is quite possible, however, that these occur at a higher horizon than the Oxford Clay of England, since some beds which are the equivalents in part at least of our Corallian, are included in the continental Oxfordian.

(iii) *To the Lower Calcareous Grit and newer beds :—*

Omitting all the doubtful species and all those which pass up from the Lower Oolites and Oxford Clay into the Corallian or higher beds, there are 25 species out of the 65 fossils from Elsworth (*i.e.* 38 per cent.), and 16 out of the 45 from St Ives (*i.e.* 35 per cent.), which are found either in the Lower Calcareous Grit or newer beds. The palæontological evidence, therefore, clearly points to a higher horizon for these limestones than the Oxford Clay itself.

(iv) *To the Lower Calcareous Grit alone :—*

33 species out of the 65 (*i.e.* 50 per cent.) from the Elsworth Rock, and 30 out of 45 (*i.e.* 66 per cent.) from the St Ives, occur in the Lower Calcareous Grit of other areas. When it is remembered that the thickness of the Lower Calcareous Grit in Yorkshire is from 80 to 100 feet, whilst the limestone at Elsworth is only 3 to 7 feet, the large percentage of fossils common to both must be regarded as linking these beds very closely together. Further, the following species occur in the Elsworth and St Ives Rocks, which are highly characteristic of, and, as yet, have only been found in the Lower Calcareous Grit:—

Waldheimia bucculenta (Sow.)

      ,,   *Hudlestoni* (Walk.)

Millericrinus echinatus (Schloth.)

In addition to these, the following species also occur which are characteristic of, but not peculiar to, the Lower Calcareous Grit:—

Ammonites perarmatus, Sow. (Found also in the upper zone of the Oxford Clay and in the Coralline Oolite).

Modiola bipartita, Sow. (Kellaways Rock, Oxford Clay and Coralline Oolite).

Collyrites bicordata (Leske) (Coralline Oolite).

The clay underlying the rock at St Ives contains *Ammonites perarmatus*, which, with the Ammonites accompanying it, indicates the highest zone of the Oxford Clay, and the beds which naturally succeed it should be the Lower Calcareous Grit.

Both the stratigraphical and palæontological evidence, therefore, show that the Elsworth and St Ives Rocks must be regarded as the equivalent in part at least of the Lower Calcareous Grit.

Fig. 2. *Section from St Ives to Bourn to illustrate Prof. Seeley's views on the position of the Elsworth and St Ives Rocks.*

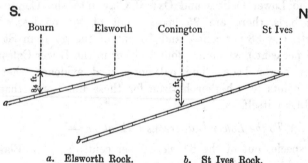

a. Elsworth Rock.   *b.* St Ives Rock.

It now remains to discuss the evidence on which Professor Seeley's views are founded as regards the relative positions of these limestones. It has already been stated that he places the Elsworth Rock in the highest zone of the Oxford Clay, and the St Ives Rock 130 feet below it.

The position of the St Ives Rock is arrived at principally on stratigraphical evidence. He assumes that (1) the rock in the well at Conington to be the same as that of St Ives; (2) the rock

in the well at Bourn to be an extension of that of Elsworth; and
(3) the dip of the beds between St Ives and Bourn to be fairly
constant. The section here given (fig. 2) illustrates these points.

If these assumptions were correct, the position assigned to the
St Ives Rock would be the true one. It has been shown, however, that the dip of the St Ives Rock itself is easterly and *not*
southerly, whilst that of Elsworth is southerly[1]. What evidence
there is then shows that the dip of these rocks is variable.
Further there is no evidence whatever to show that the rock at
Conington is the same as that of St Ives. A section taken from
St Ives through Elsworth would be somewhat as shown in the
following section (fig. 3):

FIG. 3. *Section from St Ives to Elsworth.*

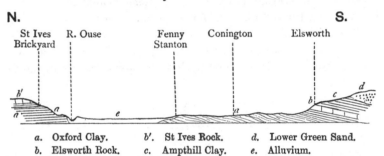

a. Oxford Clay.	*b'.* St Ives Rock.	*d.* Lower Green Sand.
b. Elsworth Rock.	*c.* Ampthill Clay.	*e.* Alluvium.

Reference has already been made to the rock in the Bluntisham
cutting, which Professor Seeley considers to belong to the Elsworth
Rock series. It is not now exposed but in the cutting further
north the overlying clays dip north at a low angle, and it is highly
probable that the dip of the rock at Bluntisham is the same. It
would seem therefore that the Elsworth and St Ives Rocks and
the one in the Bluntisham cutting, which are all on the same
horizon, are bent into a broad anticlinal, the centre of which lies
south of the St Ives brickyard. This anticlinal is inclined to the
east because these limestones were met with in the well boring at
Swavesey and Chettering Farm.

Professor Seeley admits that there is no considerable difference
in the fossils of the Elsworth and St Ives Rocks, and he also

---

[1] Professor Seeley states that the dip of the Elsworth Rock is south, and afterwards
that it is east. *Ann. Mag. Nat. Hist.*, ser. 3, vol. x. (1862), pp. 101, 102, 104, &c.

states that the lithological character of the rocks is similar. The evidence he brings forward is scarcely sufficient to establish the view of the St Ives Rock being at a lower horizon than that of Elsworth. Professor Seeley considers the possibility of their being identical and states[1] that "nothing but its simplicity is known in support of this hypothesis, which is only mentioned here as a contingency which has not been overlooked. Even should it be ultimately proved true, the result would in no way affect the principal object of this paper, except in giving the arguments greater weight, by showing that the Elsworth Rock is really lower in the Oxford Clay, than it is now supposed to be. Meanwhile I adopt the explanation previously given as the true one." It has been shown, however, that the conclusions arrived at from palæontological and stratigraphical considerations point to a higher horizon than the Oxford Clay, for these limestones.

The Elsworth Rock is placed by Professor Seeley in the upper zone of the Oxford Clay chiefly on palæontological evidence. He gives the following lists of fossils as occurring in this rock:—

(i) CEPHALOPODA.

Ammonites vertebralis, Sow.
      „   *biplex*, Sow.
      „   *perarmatus*, Sow.
      „   *Henrici*, d'Orb.
      „   *canaliculatus*, Münst.
      „   *Goliathus*, d'Orb., *var.*

Two species only of these, he states[2], "have yet been published from the Lower Calcareous Grit." It will be seen from the table given above, that all of them occur in the Lower Calcareous Grit except *A. Henrici*, and this species has already been referred to. Professor Seeley also records *Am. Rüppelli*, Münster, but there is no ammonite referable to this species now in the Woodwardian Museum collection from this rock.

Belemnites hastatus and *B. tornatilis* are said to occur in the Oxford Clay. The first is now known to occur also in the Lower Calcareous Grit. There are fragments of a large Belemnite in the

---

[1] *Ann. Mag. Nat. Hist.*, ser. 3, vol. x. (1862), p. 104.
[2] *Ibid.*, p. 108.

Woodwardian Museum, which, I believe, has been named *tornatilis* by Professor Seeley; it may belong to some other species.

(ii) Gasteropoda.

Pleurotomaria Münsteri, Rœmer
      „   *amphicœlia*, Seeley, MS.
Littorina perornata, Seeley, MS.
      „   n. sp.
Cerithium, n. sp.
Phasianella, sp.

P. *Münsteri* has been found in England only in beds higher than the Oxford Clay. I have not met with any specimen of *Phasianella* from this rock.

(iii) Lamellibranchiata.

[1] *Avicula pterosphena*, Seeley, MS.
  Gryphœa elongata, Seeley, MS.
× *Pecten fibrosus*, Sow. (including *vagans*, Sow.)
×   „   *lens*, Sow.
×   „   *vimineus*, Sow.
× *Lima pectiniformis* (Schloth.)
× *Avicula expansa*, Phil.
×   „   *ovalis*, Phil.
     „   *elliptica*
× *Trigonia costata*, Sow. (should be *T. elongata*, Sow.)
×   „   *clavellata*, Sow. var. *decorata*, Lycett, var. (this is *T. perlata*, Agassiz)
× *Astarte ovata* (Smith)
     „   *lurida*, Phil.
× *Opis Phillipsi*, Morris
× *Myacites recurva* (Phil.) var.
× *Gryphœa dilatata*, Sow.

(iv) Brachiopoda.

Terebratula ornithocephala, Sow.
     „   *perovalis*, Sow.
     „   *sphœroidalis*, Sow.

[1] *A. pteropernoides*, Blake and Hudleston.

None of these exists in the Woodwardian Museum collection. The only brachiopods which occur are given in the table (p. 27).

All the species marked × occur in beds higher than the Oxford Clay; Professor Seeley admits that most of them occur in the Coral Rag. Nevertheless, in summing up the evidence derived from the fossils, he states[1] that "when the circumstances already pointed out are remembered, and also that many Coral Rag forms range down to the Cornbrash, the fossils will be regarded as far more closely linked to the beds below than to the equivalents of strata above. When one remembers the superposition, I see no ground whatever on which the conclusion need be disputed; and therefore the Elsworth Rock will be placed as about the highest zone of the Oxfordian series." It has already been shewn, however, that the evidence derived from the fossils, and also from the superposition of the beds, is in favour of regarding the Elsworth and St Ives Rocks as Lower Calcareous Grit.

Mr Barrett, as already stated, maps the Elsworth Rock as Upper Calcareous Grit, because the overlying clays contain *Ostrea deltoidea*. It will be shewn, however, that these clays also contain some Oxford Clay forms, and are in reality a deposit intermediate between the Oxford and Kimeridge Clays. The palæontological evidence is opposed to considering the Elsworth Rock as Upper Calcareous Grit.

[1] *Op. cit.* p. 109.

V. Ampthill Clay.

THE name of this clay is taken from Ampthill in Bedfordshire, where it is well exposed. In the Cambridge district it lies conformably on what is considered to be the Lower Calcareous Grit (the Elsworth Rock). The upper limit of the Ampthill Clay is drawn at the base of a phosphatic nodule bed, which is taken as the basement bed of the Kimeridge Clay; this nodule bed occurs at Haddenham and in the well-boring at Chettering, a description of it will be given in the account of the Kimeridge Clay. If these conclusions be correct, the Ampthill Clay must have been deposited during the period when the calcareous, argillaceous and arenaceous material of the Coralline Oolite, Coral Rag and Upper Calcareous Grit was being laid down in Yorkshire and the southern counties of England. In other words the Ampthill Clay is the equivalent of all the Corallian Rocks, with the exception of a portion of the Lower Calcareous Grit.

Much of the Ampthill Clay has been mapped by the Geological Survey as Oxford Clay; it will be seen, however, that it is distinct from that formation and also from the overlying Kimeridge Clay. There is no great physical break separating the three clays, although a slight unconformity probably separates the Kimeridge and Ampthill Clays. Their lithological characters are somewhat similar. In this district therefore, "approximately uniform physical conditions prevailed throughout the Middle Oolite and the earlier part of the Upper Oolite period[1]." As a necessary result of this, it will be seen that the fauna is somewhat peculiar, since it contains a mixture of Oxford Clay, Corallian and Kimeridge Clay forms, and some which are new.

The Ampthill Clay lies to the east of the tract of country formed by the Oxford Clay; the boundary between the two cannot

[1] T. G. Bonney, *Cambridgeshire Geology* (1875), p. 15.

be well defined, because the country is largely covered by drift and exposures are few. The Ampthill Clay is worked in the brick-pits of Everton, west of Great Gransden, and north of Caxton. Further north it sweeps round to the east, and occurs east of St Ives. Beyond this I have seen sections in it, to the north of Needing-worth, in the railway cutting west of Bluntisham, and at Fenton, three miles north-west of Somersham. Its western limit would therefore be to the west of these localities.

Its eastern boundary is formed by the outcrop of the Lower Greensand at Everton, Great Gransden, Eltisley, and Papworth; it then turns easterly and is bounded by the Lower Kimeridge of Knapwell, Oakington and west of Willingham. The Ampthill Clay thus occupies the narrow band of country between the Oxford Clay and Lower Greensand from the south of Everton to the north of Papworth; to the north of this it widens out easterly and forms the low ground north of Boxworth, around Longstanton, Swavesey, Over and Willingham. It also extends further north, and finally dips under the Fenland beyond Somersham. There are three clay-pits in this formation near Gamlingay, and all of them shew a thin capping of Lower Greensand; the same thing occurs at Ever-ton. In all the other sections which I have seen, the clay is over-lain by drift, except at Knapwell, where there is a thin layer of Lower Kimeridge Clay.

The Ampthill Clay consists principally of black tenacious clays, at times almost carbonaceous. When weathered it is greyish-blue in colour. The clay usually contains crystals of selenite, and these often occur in great abundance, sometimes disseminated through the clay itself, but more especially along the joints. It also con-tains some pyrites, and very occasionally a few nodules of phos-phate of lime. Several limestone bands are found in the clay, they are usually of a greyish colour, and may be soft and somewhat sandy or very hard and compact. Some of the bands are nodular, and these nodules shew a septarian structure. Fossils are usually common, but they are often fragmentary and rarely well preserved; they are *never* pyritized, and on this account the clay is easily dis-tinguished from the underlying Oxford Clay.

The following sections have been observed in this clay :—

(1) The most southerly is in a clay-pit below the Lower

Greensand escarpment, immediately to the west of the village of Everton. About 28 feet of clay is seen in the pit. On the surface, where the clay has been exposed for some time, it presents a greyish appearance, but when fresh dug it is quite black and tenacious. Occasionally it contains a slight admixture of sand and small crystals of selenite occur throughout. About 10 feet from the bottom of the pit there is a thin band of greyish sandy lime-stone, about one foot thick; it is much divided by vertical joints. The following fossils were obtained from the clay :—

> *Belemnites abbreviatus*, Mill.
> *Ammonites cordatus*, Sow., var. *cawtonensis*,
>   Blake & Hudl.
> *Alaria bispinosa* (Phil.)
> *Trigonia paucicosta*, Lyc.
> *Arca rhomboidalis*, Contej.
> *Corbula Deshayesea*, Buv.
> *Astarte supracorallina*, d'Orb.
> *Thracia depressa* (Sow.)
> *Gryphœa dilatata*, Sow.
> *Ostrea discoidea*, Seeley
> „   *deltoidea*, Sow.

(2) The clay-pits near Gamlingay :—

(i) South of the last "*g*" of "Gamlingay Heath" is a small pit which shews about 10 feet of black clay with fragmentary fossils: it is overlain by Lower Greensand. When the clay was worked at a greater depth, two calcareous beds were pierced.

(ii) The largest clay-pit in the Gamlingay district occurs at Gamlingay Bogs, in it the following section is seen :—

							feet
(a)	Lower Greensand	.	.	.	.	.	10
(1)	Greyish-black clays	.	.	.	.	.	11
(2)	Grey argillaceous limestone		.	.	.	.	1
(3)	Black clays	.	.	.	.	.	9
(4)	Grey argillaceous limestone		.	.	.	.	1
(5)	Clay . . . 2 to 3 ft.						
(6)	Calcareous bed—not pierced						

} not now seen.

The limestone (4) is somewhat nodular, and is in great part

made up of *Serpulæ*; shells of *Ostrea* and other molluscs also occur in it. This limestone now forms the floor of the pit, but the clay below has, at one time, been worked. The following fossils occur in the clays:—

> *Belemnites abbreviatus*, Mill.
> *Ammonites biplex?*, Sow.
> *Alaria bispinosa* (Phil.)
> *Cucullœa contracta*, Phil.
> *Pecten fibrosus*, Sow.
> 　　　„　　*Thurmanni*, Contej.
> *Gryphœa dilatata*, Sow.
> *Ostrea discoidea*, Seeley

(iii)　Between Gamlingay Bogs and the village of Gamlingay, is a third clay-pit, in which occurs:—

			feet
(a)	Lower Greensand	. . . . .	3 to 4
(1)	Black clay	. . . . . .	18
(2)	Grey argillaceous limestone	. . . .	1

The clay is black when fresh dug, but on exposure it becomes greyish in colour and much jointed. It contains some nodules of pyrites and small crystals of selenite: the latter, however, are not very abundant. The following fossils were collected from the clay:—

> *Ammonites cordatus*, Sow. var. *cawtonensis*, Bl. & H.
> *Alaria bispinosa* (Phil.)
> *Nucula Menki*, Rœm.
> *Thracia depressa* (Sow.)
> *Astarte supracorallina*, d'Orb.
> 　　　„　　sp.
> *Arca*, sp.
> *Gryphœa dilatata*, Sow.
> *Ostrea discoidea*, Seeley

This pit occupies higher ground than the preceding one, and the calcareous band at its base, is the same as the uppermost one in the pit at Gamlingay Bogs. The following section would, therefore, represent the whole of the Ampthill Clay as seen near Gamlingay:—

feet

1. Black clay 18
2. Limestone 1
3. Black clay 9
4. Limestone 1
5. Black clay . . . . 2 to 3
6. Limestone . . . . . . (not pierced)

(3) About one and a half miles west of Great Gransden is a small pit, in which is seen:—

feet

(a) Soil
(b) Flint gravel
(c) Greyish clay, the upper part reassorted, and containing much selenite 7
(d) Sandy, much jointed limestone 1

The workmen informed me that, at one time, they dug much deeper than this, and below the limestone met with:—

(e) Black clay with selenite 8
(f) Calcareous bed 1
(g) Clay 3
(h) Calcareous bed . . . . . (not pierced)

Fossils:—

Gryphœa dilatata, Sow.
Ostrea discoidea, Seeley

About a mile east of this (*i.e.* half a mile west of Great Gransden), is another pit, in which is exposed:—

ft. in.

(a) Soil
(b) Clay with selenite crystals, its upper portion reassorted 6 0
(c) Bed of septarian nodules; the interior of the nodules much fissured, and the fissures filled with barytes 9
(d) Black clay 9 0

I was informed that below this the clay contains so many fragments of shells, as to render it useless for brick-making. The clay also contains small crystals of selenite and a few nodules of phosphate of lime. It yielded the following fossils:—

> *Ammonites cordatus,* Sow.
> „   *excavatus,* Sow.
> *Cardium,* sp.
> *Ostrea discoidea,* Seeley
> *Gryphœa dilatata,* Sow.

This pit is situated on higher ground than the one further west: and it seems probable that the beds exposed in it are above those of the latter. In the following section I have given what I consider to be the arrangement of the beds in these two pits, and I have also added a list of the Gamlingay beds for comparison: it will be seen that they agree very closely the one with the other, and indeed may be the same beds.

				Ampthill Clay near Gamlingay		
	Ampthill Clay west of Great Gransden.	ft.	in.		ft.	in.
(1)	Clay with selenite .	6	0			
(2)	Septarian nodule bed .	0	9			
(3a)	Black clay . . .	9	0 } ...(1)	Black clays	18	0
(3b)	Black clay with selenite	7	0			
(4)	Limestone . . .	1	0 ... (2)	Limestone .	1	0
(5)	Black clay . . .	8	0 ... (3)	Black clay .	9	0
(6)	Limestone . . .	1	0 ... (4)	Limestone .	1	0
(7)	Clay	3	0 ... (5)	Clay .	2 to 3 ft.	
(8)	Limestone . (not pierced) ... (6)			Limestone		
				(not pierced)		

I do not insist on the identity of the beds in the two localities, but simply give the comparative sections to shew that there is a possibility of their being identical. The majority of the fossils at Gamlingay came from a very fossiliferous band in the black clay (1 of the above section) and about 8 feet above the calcareous bed (2); this seems to agree very well with the description of the clays (*a*) of the pit half a mile west of Great Gransden, the clay being too full of fossils to be used below 9 feet.

(4) There is a small clay-pit about half a mile west of Croxton

church, which, at the time of my visit, only shewed 4 feet of black clay with selenite crystals. The workmen, however, informed me that the clay had been worked to a depth of 20 feet, and that two calcareous beds had been met with, one about 10 feet from the surface, and another 9 feet lower down. I saw no fossils which could be identified in the clay, but large shells (probably *Ostrea*) and *Belemnites* were sometimes found.

I was unable to find any exposures in this clay between Croxton and Elsworth. At Papworth St Everard, one and a half miles west of Elsworth, a rock was met with in a well, at a depth of 7 feet from the surface, but not pierced. Professor Seeley states[1] that this rock comes below that of Elsworth. The dip of the latter, however, is mainly to the south, and since Papworth stands on higher ground, I should be inclined to place it above rather than below the Elsworth Rock. It has been shewn that the Ampthill Clay occurs further west, at Croxton, so that it seems highly probable that the limestone at Papworth is one of the many limestone bands which occur in the Ampthill Clay.

(5) In the banks of the stream about three-quarters of a mile south of Elsworth and about half a mile west of the "K" in Knapwell, is a section about 12 feet long, which shews 3 feet of black clay full of selenite crystals and some decomposed pyrites; near its base is a bed, about a foot thick, of grey sandy limestone weathering white; it is much jointed and somewhat nodular. The following fossils were obtained from the clay and limestone:—

> *Ostrea discoidea*, Seeley
> *Gryphœa dilatata*, Sow.
> *Alaria bispinosa* (Phil.)

This clay lies some distance above the outcrop of the Elsworth Rock.

Professor Seeley states that "in tracing the brook to the south, I found in its bed three successive layers of a hard whitish-grey rock, 6 or 8 inches in thickness, occurring in the [Ampthill] clay at heights above the rock of Elsworth of about 7, 15 and 20 feet. However, these only contain a few *Gryphœa dilatata*, and

---

[1] *Ann. Mag. Nat. Hist.*, ser. 3, vol. x. (1862), pp. 99 and 100.

resemble more than anything else layers of hard, dark Lower Chalk[1]."

The section described above is the only one now visible in the Ampthill Clay near Elsworth: I have reason to believe that it occurs at a higher horizon than the highest limestone bed described by Professor Seeley and referred to above. In addition to the fossils already given from the Ampthill Clay above the Elsworth Rock near Elsworth, Professor Seeley records the following:—

> *Ammonites vertebralis*, Sow.
> *Lima pectiniformis* (Schloth.)
> *Pecten lens*, Sow.
> *Ostrea Marshi*, Sow.

(6) About a mile east-south-east of Elsworth is the village of Knapwell, and immediately to the north of the church is a brick-pit which is not now being worked. When open it shewed about 18 feet of dark blue clays; they contain aggregations of decomposed pyrites and selenite, the latter having a radiate arrangement. The crystals of selenite in the upper part of the clay are very fine, some being 8 inches in length, but usually they are only from 2 to 3 inches. At the bottom of the pit there is a light coloured limestone, weathering white and containing fragments of shells. About 3 feet below the surface, the clay contains many nodules of phosphate of lime, and this I take to be the basement bed of the Kimeridge Clay.

In the Survey Memoir[2] the whole of the clay in the Knapwell pit is included in the Kimeridge Clay. I have been unable to secure any fossils from the Knapwell pit, but *Gryphæa dilatata* occurs there.

(7) Two miles further east, and north of the village of Boxworth, is a brickyard in which two pits have been opened. The easterly one shews:—

feet

(a) Soil and drift
(b) Bluish-grey clay with fragments of shells .   . 3 to 4
(c) Limestone band which is made up of two nodular

---

[1] *Ann. Mag. Nat. Hist.*, ser. 3, vol. x. (1862), p. 100.
[2] W. H. Penning and A. J. Jukes-Browne, *Geol. Neighbourhood of Cambridge* (1881), p. 9.

feet

beds in close proximity; at times they are separated by a thin parting of clay. The limestone is greyish in colour, is very hard and compact, and occasionally has a cherty appearance . . . . 2 to 3

(d) Black fossiliferous clay with some selenite crystals . . . . 4 feet seen

A layer of septaria is said to occur several feet below the limestone[1].

The beds dip south-west at a very low angle.

The second pit lies a little to the south-west of this one and is in black clay, of which only 5 feet are now seen; it contains much selenite and some decomposed pyrites. Fossils are numerous, but are very fragmentary and badly preserved; I was unable to secure any that were identifiable. This clay comes above the bed exposed in the eastern pit.

The following fossils were obtained in the first of the two pits, those marked (l) come from the limestone:—

(l) *Ammonites plicatilis*, Sow.
(l)   ,,   *excavatus*, Sow.
      ,,   *Achilles*, d'Orb.
(l) *Nautilus hexagonus*, Sow.
(l) *Alaria bispinosa* (Phil.)
(l) *Pleurotomaria*, sp.
(l) *Nucula Menki*, Rœm.
(l) *Trigonia clavellata*, Sow.
(l) *Myacites decurtata?* (Phil.)
(l) *Plicatula*, sp.
Pecten vagans, Sow.   ⎫
  ,,   *lens*, Sow.   ⎬ Clay and
  ,,   *Thurmanni*, Contej. ⎭ Limestone
Ostrea læviuscula?, Sow.
  ,,   *discoidea*, Seeley
  ,,   *deltoidea*, Sow.
Gryphœa dilatata, Sow.
Discina Humphriesiana (Sow.)
Serpula tetragona, Sow.

[1] Penning and Jukes-Browne, *op. cit.* p. 7.

Professor Seeley describes[1] the rock in the Boxworth pit as being
1 foot 6 inches thick.   He states that "It is dark-blue, very hard,
and divided into layers, much as is the Elsworth Rock.   The only
specimen of it I saw was a slab from the upper part, about 6 inches
in thickness, which consisted of two layers—an upper dark-blue
one, with a few shells scattered about in it, and a lower pale
brown layer composed almost *entirely* of shells, chiefly uni-
valves."

I was unable to find the shelly layer referred to by Professor
Seeley, though the whole thickness of the rock is now well exposed for some distance.   It seems, therefore, that this limestone
is somewhat variable in character.   In the Survey Memoir[2], the
limestone is said to be separated into two layers by a foot of clay;
as at present seen, however, only a very thin parting of clay exists
and even this is absent in some parts of the section.   In addition
to the fossils already given from the Boxworth pit, Professor
Seeley records the following :—

Ammonites biplex, Sow.
Cerithium muricatum (Sow.)

The "Boxworth Rock" may possibly be the same as the rock in
the stream half a mile west of the "K" in Knapwell.   The fossils
Ostrea discoidea and *Gryphœa dilatata* were common in the clays
there, both above and below the limestone just as at Boxworth.
At any rate its position is in the Ampthill Clay and some distance
above the Elsworth Rock.

(8) It has already been stated that a thin layer of clay overlies the St Ives Rock in the brickyard west of St Ives.   Although
I have examined this clay on several occasions, I have, as yet,
found no fossils in it.   Professor Seeley, however, records *Ammonites alternans* and *A. Babeanus*.   It no doubt belongs to the
Ampthill Clay.   The St Ives Rock dips to the east at a low angle,
so that the clays at Swavesey (three miles south-south-east of
St Ives), which I now proceed to describe, must overlie the limestone of St Ives.

(9) During this summer (1885) a long drain, 15 feet deep, has

[1] *Ann. Mag. Nat. Hist.*, ser. 3, vol. x. (1862), p. 104.
[2] Penning and Jukes-Browne, *op. cit.* p. 7.

been dug through the southern portion of the village of Swavesey, and at its termination a well was sunk to a depth of 23 feet. The section seen in the well was as follows:—

		ft.	in.
(a)	Soil and gravel	1	6
(b)	Brown and black clays with much selenite; fossils not numerous and mostly fragmentary . .	9	0
(c)	Black clays with better preserved fossils, and with large crystals of selenite some 2 inches in length	12	6

About 5 feet down in the black clay (c) a band of isolated septarian nodules was met with. The nodules are about a foot in diameter, nearly circular in outline and slightly concave on the upper and under surfaces. Their interior is much fissured, and the fissures were wide, but are now filled with crystals of barytes. In several nodules these crystals were seen to have a radiate arrangement, and those which were nearest the centre of radiation are pink coloured, whilst those further away are white. Some of the nodules too, have the fissures lined by a thin layer of pyrites. The calcareous material in the nodule itself is of a greyish colour, and is very hard and compact, the outer portion being somewhat softer and coloured by oxide of iron.

The following fossils were obtained from the black clays:—

> *Belemnites nitidus,* Dollf.
> [1]*Ammonites plicatilis?,* Sow.
>   „   *vertebralis,* Sow.
> [1]   „   *biplex?,* Sow.
> *Nucula Menki,* Rœm.
> *Leda,* sp.
> *Arca longipunctata,* Blake
> *Cucullœa contracta,* Phil.
> *Lucina aliena* (Phil.)
> *Thracia depressa* (Sow.)
> *Plicatula,* sp.
> *Pentacrinus,* sp.

---

[1] These fossils have merely a portion of one whorl preserved and this is much distorted.

In the northern part of the village a well was sunk through 23 feet of clay, when a hard rock was reached and excavated to a depth of 2 or 3 feet; this rock, as has already been stated, I consider to be a continuation of that of St Ives. The land rises slightly towards the south, so the clay in the well and drain described above, would lie just above this rock.

(10) In the railway-cutting, one mile east of Swavesey station, is a small exposure in black clays, containing much selenite, with *Ostrea discoidea*, *Exogyra nana*, &c.

(11) A well, 10 feet deep, has recently (June, 1885) been dug, a little to the north-east of "74" on the Survey Map and north of Needingworth. It shewed black clays with selenite crystals, and near its base was a layer of septarian nodules precisely like those which occur at Swavesey.

(12) Black clay is seen in the railway-cutting immediately south-west of Bluntisham station; in the floor of the western end of the cutting, a thin greyish sandy limestone crops out. The clay yielded *Gryphœa dilatata* and *Ostrea discoidea*.

(13) A little further to the north-west, about a mile and a half west of Bluntisham, there is a long cutting on the Somersham and St Ives railway. This cutting is about 40 feet deep, and its sides are made up in great part of Boulder Clay; but below there is some black clay with selenite crystals. Very little of it is now seen as the slopes of the cutting are overgrown with vegetation. Reference has already been made to the occurrence of a limestone like the Elsworth Rock, at the bottom of this cutting. Professor Seeley gives the following fossils as coming from the clay:—

Belemnites excentricus, Blainv.
Ammonites alternans, Von Buch
     „   *biplex*, Sow.
     „   *serratus*, Sow.
     „   sp.[1]
Gryphœa dilatata, Sow.
Ostrea deltoidea, Sow.

[1] Figured by d'Orbigny as the female of *Duncani*.

I also found in it :—

> *Pecten lens*, Sow.
> *Cerithium*, sp.

(14) At Fenton, three miles north-west of Somersham, are two brick-pits, one on each side of the road. The westerly one is in black clays, sometimes grey and brown, especially near the surface, with many small crystals of selenite and some decomposed pyrites. The fossils are not well preserved, but I obtained the following species :—

> *Belemnites abbreviatus*, Mill.
> „ *hastatus*, Blainv.
> *Ammonites cordatus*, Sow. var. *cawtonensis*, Blake & Hudl.
> „ *vertebralis*, Sow.
> *Cucullœa contracta*, Phil.
> *Thracia depressa* (Sow.)
> *Ostrea discoidea*, Seeley

The pit on the east side of the road, is now abandoned ; in one portion of it, however, a grey sandy limestone, about 6 inches thick, was seen, and I was informed by the workmen that another bed of limestone had been met with lower down.

(15) In the well-boring at Chettering (see p. 23) there was pierced 26 feet of black clay which is probably the Ampthill Clay. It overlies what I consider to be the Lower Calcareous Grit and is overlain by a bed of clay rich in phosphatic nodules, which I take to be the basement bed of the Kimeridge Clay.

The following well-sections are taken from Appendix C, of the Geological Survey Memoir on sheet 51 S.W. :—

COTTENHAM (p. 160).

	feet
Black earth	4
Blue clay mixed with rock	200
Rock and sand	

The greater part of the blue clay is of Ampthill Clay age, though its upper portion is probably Kimeridgian. The 'rock and sand' may be the Lower Calcareous Grit, and appears to be the same as that met with in the Chettering boring.

OVER (p. 163).

	feet
Black earth and gravel	15
Blue (Oxford) clay with chalk stone and black rocks .	200

The 'Blue clay' is Ampthill and not Oxford Clay. The railway-cutting one mile east of Swavesey, already referred to, contains Ampthill Clay, and this lies but a short distance south of Over. The Lower Calcareous Grit does not appear to have been reached in this boring.

RAMPTON (p. 168).

		feet
Oxford Clay	Solid brown clay with a white stone . . .	4
	White rock and gravel, very hard . . .	2
	Coloured clay	2
	Blue clay streaked with white and small spots of talc (selenite)	2

This clay, again, is Ampthill. In the Woodwardian Museum the following fossils are labelled as coming from this locality:—

Nucula Menki, Rœm.
Thracia depressa (Sow.)
Cidaris Smithi, Wright

The above sections are the principal ones which occur in the Ampthill Clay of the district under consideration. It will be seen that they are somewhat isolated and the thickness of the strata exposed in each is not great. On this account it is not easy to estimate the thickness of the Ampthill Clay. It can, however, be approximately ascertained in the well-borings. Thus, in the boring at Over, if all the clays met with there, be of Ampthill Clay age, and supposing the Lower Calcareous Grit is not reached, which if it had been pierced would have been recorded, then the thickness of the Ampthill Clay must be at least 200 feet. The boring at Cottenham shews much the same results, except that here the Lower Calcareous Grit was reached, and the Ampthill Clay is certainly thinner than at Over. In the Chettering boring this clay is only

26 feet thick. If these conclusions be correct, the Ampthill Clay must thin out rapidly in an easterly direction, and may possibly disappear altogether before Upware is reached, where it would be replaced by the Corallian limestones.

Subjoined is a list of the fossils from the Ampthill Clay. I have indicated in it those which occur also in the Oxford Clay, Corallian Rocks and Kimeridge Clay. It will be seen, from this list, that the fossils are a mixture of forms from all three formations. Out of the 41 species determined 30 occur in the Corallian and 15 and 19 in the Oxford and Kimeridge Clay respectively. The fossils of the Ampthill Clay, therefore, link it more closely to the Corallian than to either the underlying or overlying formations. Mr Jukes-Browne states[1] that the Ampthill Clays "do not contain a Corallian fauna, only a mixture of Oxford and Kimeridge Clay species." The accompanying list, however, shews that the fauna of the Ampthill Clay is most nearly allied to the Corallian, and that the Oxford and Kimeridge species it contains are principally those which pass up into the Corallian or pass up from that formation into the Kimeridge Clay in other areas.

The only fossils peculiar to the Ampthill Clay are the saurians and *Ostrea discoidea*. *Ostrea deltoidea* and *Gryphœa dilatata* occur together in these beds, and also in the Corallian, but in no other formation.

LIST OF FOSSILS FROM THE AMPTHILL CLAY OF THE NEIGHBOURHOOD OF CAMBRIDGE.

S indicates that the species is quoted on the authority of Professor Seeley.

Geological Distribution

	Oxford Clay	Corallian	Kimeridge Clay and Passage Beds
REPTILIA			
S *Pliosaurus pachydeirus*, Seeley			
S *Cryptodraco eumerus* (Seeley)			
CEPHALOPODA			
Belemnites abbreviatus, Mill.		×	×
" *nitidus*, Dollf.		×	×
S " *excentricus*, Blainv.			
Nautilus hexagonus, Sow.	×	×	!
Ammonites Achilles, d'Orb.	×	×	

[1] *Historical Geology*, (1886), p. 352.

R. E. 4

List of Fossils (*continued*).

	Geological Distribution		
	Oxford Clay	Corallian	Kimeridge Clay and Passage Beds
S Ammonites alternans, Von Buch		×	×
S " Babeanus, d'Orb.	×		
S " biplex, Sow.			×
" cordatus, Sow.	×	×	×
" " var. cawtonensis, Bl. and H.		×	
" excavatus, Sow.	×	×	
" plicatilis, Sow.		×	
S " serratus, Sow.		×	×
" vertebralis, Sow.	×	×	
GASTEROPODA			
S Cerithium muricatum (Sow.)		×	
Alaria bispinosa (Phil.)		×	
Pleurotomaria, sp.			
LAMELLIBRANCHIATA			
Nucula Menki, Rœm.		×	×
Leda, sp.			
Cardium, sp.			
Trigonia clavellata, Sow.	×	×	
" paucicosta, Lyc.	×		
Corbula Deshayesea, Buv.		×	×
Cucullœa contracta, Phil.		×	
Arca longipunctata, Blake			×
" rhomboidalis, Contej.		?	×
" sp.			
Lucina aliena (Phil.)		×	
Astarte supracorallina, d'Orb.		×	×
Myacites decurtata (Phil.)	×	×	
Pecten fibrosus, Sow.	×	×	
" lens, Sow.	×	×	
" Thurmanni, Contej.			×
S Lima pectiniformis (Schloth.)		×	×
Thracia depressa (Sow.)	×	×	×
Plicatula, sp.			
Gryphœa dilatata, Sow.	×	×	
Exogyra nana (Sow.)	×	×	×
Ostrea discoidea, Seeley			
" deltoidea, Sow.		×	×
S " Marshi, Sow.	×		
" lœviuscula ?, Sow.			
BRACHIOPODA			
Discina Humphriesiana (Sow.)		×	×
VERMES			
Serpula tetragona, Sow.		×	×
ECHINODERMATA			
Pentacrinus, sp.			
Cidaris Smithi, Wright		×	×

VI. THE CORALLIAN ROCKS OF UPWARE.

To the north of Upware, about ten miles north of Cambridge, a local, and, to some extent, a peculiar development of the Corallian limestone occurs. It has been described by Sedgwick[1], Fitton[2], Bonney[3], and by Blake and Hudleston[4].

The limestone forms a low ridge rising about 20 feet out of the flat fen-land and extending some three miles north of Upware. There are two quarries opened in this ridge, the southerly being about half a mile north of Upware, whilst the other lies about one and a half miles further north. A considerable quantity of the rock has been quarried and used for building up the banks of the Cam. The shelving ground on both sides of the ridge is partly formed of Lower Greensand, which has been largely worked for the nodules of phosphate of lime contained in it. The workings, however, are now abandoned, but several sections have been published by Mr H. Keeping, Mr J. F. Walker, and Professor Bonney, which shew the relation of the overlying beds to the Corallian limestones below. A full account of these is given[5] in the Sedgwick Essay for 1879. The following is a summary of the published sections from the Lower Greensand pits on the western side of the ridge:—

[1] *Rep. Brit. Assoc.* 1845 (1846), Sections, p. 40.
[2] *Trans. Geol. Soc.*, ser. 2, vol. IV. (1836), p. 307.
[3] *Cambridgeshire Geology* (1875), p. 16.
[4] *Quart. Journ. Geol. Soc.*, vol. XXXIII. (1877), p. 313.
[5] W. Keeping, *The Fossils and Palæontological Affinities of the Neocomian Deposits of Upware and Brickhill*, (1883), p. 3 *et seq*.

(*a*) Soil
(*b*) Lower Gault with phosphatic nodules at its
      base . . . . 2 to 8 ft.
(*c*) Lower Greensand with an upper and lower
      phosphate bed . . . . . about 8 ft.
(*d*) Kimeridge Clay, which near its base is mixed
      with a quantity of broken fragments from the
      Coral Rag . . . . thickness
                                 not given.
(*e*) Corallian limestones.

A section on the east side of the ridge, near Spinney Abbey
Farm, is also given[1], and the beds overlying the Corallian there
are much the same as those on the west side of the ridge.

Messrs Blake and Hudleston figure[2] the Lower Greensand as
overlying the ridge of limestone; this, however, is incorrect, as the
sands are limited to the sides of the ridge.

At the entrance from the river Cam into the south quarry, the
lower portion of the Gault, with its phosphatic nodules, can be seen
in close proximity to the Coral Rag.

The limestone exposed in the two quarries differs both in its
lithological and palæontological characters. In the south quarry
about 20 feet of creamy white irregular limestone is seen. It is to
a large extent composed of corals belonging to the genera *Tham-*
nastræa, Isastræa, Rhabdophyllia and *Montlivaltia.* The spread-
ing coral *Thamnastræa arachnoides* forms thin irregular layers of
compact limestone, which is now crystalline; between these layers
the rock is usually much softer, more earthy and frequently oolitic,
and in this the majority of the fossils occur. The corals are frequently found bored by *Lithodomi.* As a rule the limestone is
pure, but a slight admixture of clay occasionally occurs, and some
of it may now be seen in the middle of the eastern bank of the pit.
The beds dip to the west of north at about 4°.

From the character of the rock it seems probable that this was
once the site of a coral reef, since the whole of the limestone is
formed of the hard parts of corals and their detritus.

---

[1] W. Keeping, *The Fossils and Palæontological Affinities of the Neocomian
Deposits of Upware and Brickhill*, (1883), p. 7.
   [2] *Quart. Journ. Geol. Soc.*, vol. XXXIII. (1877), p. 315.

Fossils are abundant, but the majority are in the form of casts. The subjoined list contains the species which have come from the south quarry.

LIST OF FOSSILS FROM THE SOUTH QUARRY, UPWARE[1].

CEPHALOPODA.

Ammonites Achilles, d'Orb.
(a)   ,,   *mutabilis*, Sow.
    ,,   *plicatilis*, Sow.
    ,,   *vertebralis*, Sow. var. *cawtonensis*, Bl. and H.

GASTEROPODA.

Natica Clymenia, d'Orb.
    ,,   *Clytia*, d'Orb.
Chemnitzia heddingtonensis (Sow.)
(a) *Pseudomelania striata* (Sow.)
Cerithium muricatum (Sow.)
Alaria, sp.
Littorina muricata (Sow.)
Neritopsis decussata (Münst.)
(a)   ,,   *Guerrei*, Héb. and Desl.
Amberleya princeps (Rœm.)
Trochus, sp.
Pleurotomaria reticulata (Sow.)
    ,,   sp.
Trochotoma tornatilis (Phil.)
Emarginula Goldfussi, Rœm.
Fissurella corallensis, Buv.

LAMELLIBRANCHIATA.

Gastrochœna Moreana, Buv.
Goniomya v-scripta (Sow.)
(a) *Myacites decurtata* (Goldf.)
    ,,   *recurva* (Phil.)
    ,,   *Voltzi* (Ag.)
Pholadomya decemcostata, Rœm.

---

[1] The list of Upware fossils, contributed by the author to the Survey Memoir on *Parts of Cambridgeshire and of Suffolk* (1891), was taken without alteration from the manuscript of this essay, which was written in August 1885. Subsequently the author made a few changes in the names attached to the specimens in the Woodwardian Museum, and the list now given has been modified accordingly. A few of the species here recorded (marked *a*) are not to be found in the Museum, some of these appear to be taken from the list given by Messrs Blake and Hudleston.—*H. W.*

Homomya tremula (Buv.)
Quenstedtia lævigata (Phil.)
       ,,   ,,   var. *gibbosa* ?, Hudl.
(*a*) *Astarte aytonensis*, Lyc.
    ,,   *ovata* (Smith)
    ,,   sp.
Cypricardia glabra, Blake and Hudl.
Opis arduennensis, d'Orb.
    ,, *corallina*, Damon
    ,, *Phillipsi*, Morr.
    ,, *virdunensis*, Buv.
    ,, sp. (? *paradoxa*, Buv.)
Myoconcha Sæmanni, Dollf.
(*a*)   ,,   *texta* (Buv.)
(*a*) *Lucina globosa*, Buv.
    ,, *Moreana*, Buv.
Cardium, cf. *delibatum*, De Lor.
Cardita ovalis, Quenst.
Trigonia Meriani, Ag.
Arca æmula, Phil.
    ,, *anomala*, Blake and Hudl.
    ,, *contracta* (Phil.)
    ,, *pectinata* (Phil.)
    ,, *quadrisulcata*, Sow.
Cucullæa elongata, Phil. *non* Sow.
Isoarca texata, Münst.
    ,, *multistriata*, Etal.
Mytilus pectinatus, Sow.
    ,, *rauracicus*, Greppin
    ,, *ungulatus*, Young and Bird
Modiola bipartita, Sow.
    ,, [*Lithodomus*] *inclusa*, Phil.
    ,, *subæquiplicata*, Rœm.
    ,, sp.
Pinna lanceolata, Sow.
Gervillia angustata, Rœm.
    ,, *aviculoides*, Sow.
Perna subplana, Etal.
(*a*) *Lima elliptica*, Whit.
    ,, *gibbosa*, Sow.
    ,, *læviuscula* (Sow.)

Lima rigida (Sow.)
  ,,   *rudis*, Sow.
  ,,   sp.
Pecten articulatus (Schloth.)
  ,,   *fibrosus*, Sow. (? south quarry)
  ,,   *inæquicostatus*, Phil.
  ,,   *vimineus*, Sow.
Hinnites velatus (Goldf.)
  ,,   sp. (cf. *corallina*, Hudl.)
(a) *Anomia suprajurensis*, Buv.
Plicatula fistulosa, Morr. and Lyc.
Exogyra nana (Sow.)
Ostrea gregaria, Sow.
(a)   ,,   *solitaria*, Sow.

BRACHIOPODA.

Terebratula insignis, Schübler, var. *maltonensis*, Oppel
  ,,   small sp.
Rhynchonella, small sp.

CRUSTACEA[1].

Gastrosacus Wetzleri, Meyer
Glyphea Münsteri? (Voltz)
Prosopon marginatum, Meyer

VERMES.

Serpula deplexa?, Phil.
  ,,   *tetragona*, Sow.
Vermicularia, sp.

ECHINODERMATA.

Apiocrinus polycyphus, Merian
Pentacrinus, sp.
Millericrinus, sp.
Cidaris florigemma, Phil.
  ,,   *Smithi*, Wright
Hemicidaris intermedia (Flem.)
Stomechinus gyratus (Ag.)
× *Collyrites bicordata* (Leske)
× *Holectypus depressus* (Leske)
× *Pygaster umbrella*, Ag.
× *Echinobrissus scutatus* (Lam.)
× *Hyboclypus*, n. sp.

[1] Named by Mr James Carter, F.R.C.S.

ACTINOZOA.

Stylina tubulifera (Phil.)

Montlivaltia dispar (Phil.)

Isastræa explanata (Goldf.)

Rhabdophyllia Phillipsi, Edw. and Haime

Thamnastræa arachnoides (Park.)

   ,, *concinna* (Goldf.)

PORIFERA.

Scyphia, sp.

Those marked × in the above list, come from the floor of the middle portion of the quarry and occur in a rock which is less coralline than the limestone which comes above it. It contains no spines of *Cidaris florigemma*, which are so common in the overlying beds, and its character is much more like the limestones of the north than those of the south quarry.

The northern quarry is opened in rocks of quite a different character to those of the southern. Its upper part is composed of soft oolitic limestone, with some hard portions, shewing little or no indications of bedding. Lower down there are regular beds of coarsely oolitic and sometimes pisolitic limestone, separated by softer beds similar to those which come above. The beds dip north at a low angle. The fossils are less numerous than in the south quarry; the following have been obtained:—

LIST OF FOSSILS FROM THE NORTH QUARRY, UPWARE[1].

CEPHALOPODA.

Belemnites abbreviatus, Mill.

Ammonites Achilles, d'Orb.

   ,, *plicatilis*, Sow.

   ,, *perarmatus*, Sow.

   ,, *trifidus*, Sow.

GASTEROPODA.

(*a*) *Littorina muricata* (Sow.)

   ,, *Meriani* (Goldf.)

Pleurotomaria, sp. (cast)

LAMELLIBRANCHIATA.

(*a*) *Opis Phillipsi*, Morris

Isoarca texata, Münst.

---

[1] See footnote on p. 53.

Mytilus jurensis, Mer.
  „   *ungulatus,* Young and Bird
Modiola bipartita, Sow.
(*a*) *Gervillia aviculoides* (Sow.)
Pecten fibrosus, Sow.

ECHINOIDEA.

Hyboclypus gibberulus, Ag.
Pygaster umbrella, Ag.
[1] *Holectypus depressus* (Leske)
[1] *Echinobrissus scutatus* (Lam.)
Collyrites bicordata (Leske).
Pseudodiadema versipora (Woodw.)

In addition to the quarries already described, two small pits have been opened, at about a quarter and half a mile south of the northern quarry, on the west side of the road joining the two quarries. The limestone exposed in these pits is similar to the soft oolitic rock forming the upper part of the quarry to the north of them.

Several descriptions have been given respecting the arrangement of the Corallian Rocks in the Upware ridge. Fitton figures[2] the limestone as forming an anticlinal arch running north and south and rising up through newer beds (Kimeridge Clay). Sedgwick gives[3] a similar arrangement of the beds in his section from Ely to Reach. Messrs Blake and Hudleston represent[4] the beds of the south quarry as dipping north, and those of the north quarry as dipping south, and the whole as forming a synclinal, the hollow being filled with Lower Greensand. In the text of their paper they also state that the beds in the north quarry dip south. Professor Bonney in describing these rocks states that the beds in both quarries dip north, and this was again[5] pointed out after the publication of Blake and Hudleston's paper.

The various horizons to which the limestones near Upware have been referred have already been given. I am not aware on what grounds the "Upware Rock" was placed above the "clay equivalent to the Coral Rag" by Professor Seeley. Professor

---

[1] Very common in upper portion of pit.
[2] *Trans. Geol. Soc.,* ser. 2, vol. IV. (1836), pl. xa. fig. 24.
[3] *Supplement* (1861), p. 22, Section 1.
[4] *Quart. Journ. Geol. Soc.,* vol. XXXIII. (1877), p. 315.
[5] *Geol. Mag.,* dec. 2, vol. IV. (1877), p. 476.

Bonney refers the whole of the Upware limestones, on palæontological grounds, to the true Coral Rag, as the term was then
understood.   Messrs Blake and Hudleston, from palæontological
considerations, and from a comparison with similar deposits in
other areas, place the limestones of the south quarry in the Coral
Rag, and those of the north quarry in the underlying Coralline
Oolite.   During the last couple of years the quarrying has been
carried on mainly in the floor of the middle portion of the south
pit, so that lower beds are now being exposed; these contain but
few corals, and spines of *Cidaris*, as already stated, are either very
rare or absent.   On the other hand, *Echinobrissus scutatus*, and
Holectypus depressus, which were formerly so rare in this quarry,
are fairly common in these lower beds.   On one occasion no less
than five specimens of these echinoids were found during a couple
of hours' work, and one rarely visits the quarry now without finding
one or more of them.   In addition to the echinoids already given,
the following species occur in these lower beds, which are also
found in the Coralline Oolite of the north quarry:—

> *Ammonites plicatalis*, Sow.
> *Mytilus ungulatus*, Young and Bird
> *Pygaster umbrella*, Ag.

FIG. 4.   *Section to explain the structure of the Upware ridge by means
of folds.*

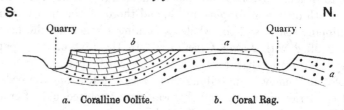

a.   Coralline Oolite.   b.   Coral Rag.

FIG. 5.   *Section to explain the structure of the Upware ridge by means
of a fault.*

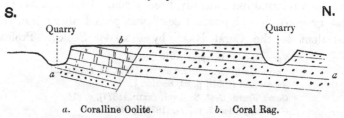

a.   Coralline Oolite.   b.   Coral Rag.

It seems highly probable, therefore, that the Coralline Oolite exists in the floor of the south pit, and comes below the true Coral Rag; thus confirming Blake and Hudleston's view as to the position of the beds. This being the case, one of two things must take place; either the beds rise soon after leaving the south quarry and then bend down again before reaching the northern, as illustrated in the section (fig. 4), or, an east and west fault, with a downthrow to the south exists somewhere between the two quarries, as shewn in the section (fig. 5).

Only the two small pits referred to are opened between the two quarries, and no sections exist in the district immediately adjoining the ridge: so that it is impossible to say which of the above explanations is the true one.

As far as is now known, the Upware ridge is the only place where the Corallian limestones are known to exist in the Cambridge district. Dr Fitton states, on the authority of Professor Sedgwick, that the Coral Rag crops out beyond Haddenham, going towards Chatteris. I have worked the Haddenham district, but was unable to find any trace of it.

In the well-boring at Chettering Farm, two and a half miles north-west of Upware, the Oxford Clay was reached without passing through any limestone similar to that at Upware.

The following section is given[1] at Dimmock's Cote, Stretham Fen (about half a mile south-west of the northern quarry at Upware):—

		feet
Black earth	. . . .	3
Peat	. . . .	18
Blue (Kimeridge) clay	. . . . .	110
Rock and sand	. . . . . .	10

The " rock and sand " is probably the same as that which was met with in the Chettering boring: and if the Corallian limestones existed they would have been pierced before the " rock and sand " was reached. It appears therefore that the Upware limestones do not extend further west than the course of the river Cam.

---

[1] W. H. Penning and A. J. Jukes-Browne, *Geol. Neighbourhood of Cambridge*, (1881), p. 165.

Further east, at Wicken, the following well section is given[1]:—

	feet
Soil	3
Blue clay	10
Black rock	1
Blue clay	2
Black rock	1
Blue clay	5
Similar alternations for	178
	200

Here again the Upware limestone, if it existed, would have been met with before this depth was reached.

Professor Sedgwick states that several wells have been sunk, between Cambridge and Lynn, through the Kimeridge Clay, and the Oxford Clay was reached without meeting any Corallian.

It seems, therefore, from these considerations, that the Upware limestones occur only in the ridge north of Upware, and that it forms an isolated reef in the midst of the great "Fen-clay" formation.

Pebbles of the limestone occur in the drift of Cambridgeshire and Huntingdonshire as well as in the Lower Greensand of Upware. Its distribution may have been more extensive than what it is at present; in any case it has undergone a considerable amount of denudation to supply the pebbles referred to above.

The limestones at Upware have never been pierced, so that nothing is known of the strata which underlie the beds exposed in the northern quarry. It has been shewn that what appears to be the Lower Calcareous Grit has been met with at Dimmock's Cote, about half a mile further west, and it is highly probably that this "rock and sand" is continued below the Coralline Oolite of the Upware ridge; nothing, however, is known of what exists, or if anything does occur, between them.

[1] W. H. Penning and A. J. Jukes-Browne, *Geol. Neighbourhood of Cambridge*, (1881), p. 167.

VII. KIMERIDGE CLAY.

THE distribution of the Kimeridge Clay in the neighbourhood of Cambridge is shewn on the Survey maps, sheets 51 N.W. and 51 S.W. I have not seen any exposures in it to the south or west of Knapwell. A narrow band of the clay runs from Knapwell, through Boxworth, Oakington and on to Cottenham; it then spreads out for some miles to the north and east, and underlies the large tract of fen-country of North Cambridgeshire. Near Ely the Kimeridge Clay is overlain unconformably by the Lower Greensand, and several outliers of these sands occur to the west and south-west of Ely, forming low hills in the fens.

The Kimeridge Clay is formed of dark and bluish-black clays, which are often laminated and occasionally quite bituminous. Sometimes these clays are very arenaceous, and contain small crystals of selenite. Several beds of grey-argillaceous limestone occur in the clay; they do not exceed a foot in thickness, and may be either in the form of regularly bedded limestones, or as layers of interrupted septarian nodules. The nodules are ellipsoidal in shape, and their longest diameter varies from one to three feet; they are composed of grey compact limestone; with their interior much fissured, and the fissures filled or lined with calcite. Fossils are fairly common throughout the clay, but they are not, in all cases, well preserved. Fragments of fossil wood are not unfrequently met with.

The sections in this clay are not very numerous: the following are the principal ones which I have observed.

It has already been stated (p. 42) that the upper portion of the clay in the Knapwell clay-pit contains some phosphatic nodules, and this I consider to be the basement bed of the Kimeridge

Clay. No fossils were collected from this nodule-bed. A short distance further south, I saw a quantity of clay that had recently been thrown out of a pond, and this clay contained numerous black nodules of phosphate of lime. This is probably on the same horizon as the top-bed of the Knapwell clay-pit. The following fossils were collected from this clay :—

> *Belemnites nitidus*, Dollf.
> *Nucula Menki*, Rœm.
> *Avicula œdilignensis*, Blake
> *Exogyra nana* (Sow.)
> *Ostrea deltoidea*, Sow. (abundant).

These phosphatic nodules are referred to in the Survey Memoir[1], and it is stated that their profusion indicates "a denudation of the clay resulting in the formation of a 'coprolite bed' previous to the deposition of the Boulder Clay." Although these nodules were not seen in place, it will be noticed later on that similar nodules occur in such a position that no denudation, such as is stated above, could have taken place.

To the south of Knapwell, at "Boxworth Thoroughfares," *Ostrea deltoidea* has been found in a clay which is probably Kimeridgian.

Phosphatic nodules similar to those occurring at Knapwell[2] "are found near where the junction line of Kimeridge Clay and Lower Greensand crosses the Huntingdon Road, and west of Oakington, shewing that a layer of coprolites exists, wholly or in part, between the two formations." The position of these nodules can hardly be the same as those of Knapwell, as there the outcrop of the Lower Greensand is at least three-quarters of a mile south of the pond in which the nodules were seen.

On the west side of the "ancient Bridleway," north of Balsar's Hill, near Willingham, a black clay is exposed in some of the ditches, which contains black phosphatic nodules similar to those of Knapwell.

A short distance north of the second "d" of Haddenham Fen, south-east of Aldreth, a pond has recently been dug in which the following section was observed :—

---

[1] W. H. Penning and A. J. Jukes-Browne, *Geol. Neighbourhood of Cambridge*, (1881), p. 9.

[2] *Ibid.* p. 9.

		ft.	in.
(a)	Soil	1	0
(b)	Grey clay	5—6	
(c)	Ferruginous and seleniferous clay	1	0
(d)	Black clay with *Ostrea deltoidea* in abundance .	1	6
(e)	Clay with black phosphatic nodules . . .	1	0

Immediately to the north of Aldreth, and occupying higher ground than the pond just described, a reservoir has been dug in the Kimeridge Clay, where 9 feet of black clay with a septarian nodule band, was seen. This is distinctly above the phosphate bed.

Half a mile west of Haddenham Station, and south of the second "h" of 'Haddenham Field,' is a large brick-pit, in which the following section is seen :—

		ft.	in.
(a)	Soil	1	6
(1)	Ferruginous clay	2	0
(2)	Mottled clays with some calcareous nodules .	4	0
(3)	Thin limestone beds, greyish in colour but weathering white; it is very hard and compact and has a cherty appearance . . .	1	3
(4)	Black tenacious clays with few fossils . .	2	0
(5)	Grey nodular limestone with *Ammonites mutabilis* and some small gasteropods . . .		9
(6)	Black clay, of which only 4 feet was visible at the time of my visit. Below this, however, came black clays crowded with very large *Ostrea deltoidea*: this was underlain by clay containing a considerable quantity of black phosphatic nodules . . . . about	9	0

The following fossils were obtained from this pit :—

> *Belemnites nitidus*, Dollf.
>   ,,   *abbreviatus*, Mill.
> *Ammonites biplex*, Sow.
> *Alaria trifida* (Phil.)
> *Pleuromya donacina*, Ag.
> *Nucula Menki*, Rœm.
> *Avicula œdilignensis*, Blake
>   ,,   *dorsetensis*, Blake

Anomia Dollfusi, Blake
Exogyra nana (Sow.)
Ostrea deltoidea, Sow. (large and abundant)
   ,,   *solitaria,* Sow.
   ,,   *læviuscula,* Sow.
   ,,   *monsbeliardensis,* Contej.
Terebratula Gesneri, Etal.
Serpula intestinalis, Phil.
   ,,   *tetragona,* Sow.
Cidaris, sp.

The lower portion of this section is identical with that seen in Haddenham Fen, and here the denudation which caused the formation of the phosphate bed must have taken place at the commencement of the Kimeridge Clay period.

This basement bed of the Kimeridge Clay, with its numerous phosphatic nodules, is fairly constant in this district, and with the clays crowded with *Ostrea deltoidea* overlying it, marks an easily recognised horizon. Similar beds occur at the base of the Kimeridge Clay near Oxford.

Immediately south of Haddenham Station another clay-pit occurs. It occupies higher ground than that further west, and is opened in clays belonging to a somewhat higher horizon. The section seen is as follows:—

		ft.	in.
(a)	Soil	2	0
(1)	Black and brown clays with small crystals of selenite	7	0
(2)	Septarian nodule bed		9
(3)	Black clays	2 ft.	seen

I was informed that 4 feet lower down, a limestone bed occurs, which, as yet, had not been pierced.

A second pit is opened about 50 yards further south, where 12 feet of black clay with selenite occurs above the septarian nodule bed (2). In both pits the beds dip south at a low angle. The following fossils were obtained from these two pits:—

     Ammonites mutabilis, Sow.
     Astarte supracorallina, d'Orb.
       ,,   sp.

Thracia depressa (Sow.)
Arca rhomboidalis, Contej.
    ,,   *mosensis*, Buv.
Pecten Grenieri, Contej.
Exogyra virgula (Defr.)
    ,,   *nana* (Sow.)
Ostrea læviuscula, Sow.
Discina, small sp.
Pollicipes Hausmanni, Koch and Dunker.

The clays in these two pits are certainly higher in the series than those in the pit further west; it is quite possible, however, that the bottom clays of the former may be the same as those forming the summit of the latter.

About one mile west of Stretham, on the north side of the road leading from that village to Wilburton, a small exposure occurs in the Kimeridge Clay with *Exogyra virgula.* A similar section is seen in the pond to the north of the road.

In the well-boring at Chettering Farm, the section of which has already been given (p. 23), about 51 feet of what I consider to be Kimeridge Clay was pierced. The upper 36 feet was made up of black clay with several thin beds of greyish limestone, and below it came 12 feet of laminated clays with *Ammonites alternans;* this was underlain by a bed of clay 3 feet thick, through which were disseminated black phosphatic nodules. The latter is probably on the same horizon as the nodule bed which has been described to the west of Haddenham, and this, as I have already stated, forms the base of the Kimeridge Clay.

The hilly ground on which the village of Stuntney stands is partly formed of Kimeridge Clay. An attempt was recently made to work the phosphatic nodules in the outlier of Lower Greensand, which occurs near Stuntney; a black clay was met with below these sands. The workings are now abandoned, and the only section which can be seen is in the railway cutting to the south of the village: and near its base is a black clay with small crystals of selenite and a few fragmentary fossils. This clay is overlain by a considerable thickness of drift.

The most important section in the Kimeridge Clay of this district is found in the pit at Roslyn or Roswell Hill, about one mile north-east of Ely. The clay has been extensively worked for building up and repairing the banks of the Cam and its tributaries. It is seen on the northern and southern sides of the pit; the greater part of the western side is formed of drift and Cretaceous beds, respecting which there has been some dispute; some considering that these beds had been faulted in, whilst others maintain that the Cretaceous beds are only a huge boulder-like fragment dropped into a valley, which had been excavated in the Kimeridge Clay. The question relating to these beds, however, is outside the subject of this essay, and it is only necessary that a brief reference should be made to it.

The best section of the Kimeridge Clay is seen on the northern side of the pit, and here the beds dip gently towards the west. The following section is taken from the western extremity of the north side of the pit:—

		ft.	in.
(a)	Soil and reassorted clay	5	0
(1)	Bituminous papery shale with *Discina latissima*	7	0
(2)	Calcareous clay, with interrupted septarian nodules		7
(3)	Greyish-black shale with some thin beds of more sandy clay; near its base it is crowded with *Exogyra virgula*	3	0
(4)	Dark-grey shales with *Am. alternans*, &c. .	1	6
(5)	Thin ferruginous layer		2
(6)	Greyish shale with three sandy layers . .	5	0
(7)	Greyish sandy shale with *Trigonellites* . .	3	0
(8)	Sandy clay with interrupted septarian nodules	1	0
(9)	Bluish clay with fragmentary fossils . .	1	6
(10)	Papery shale somewhat sandy and crowded with fossils, passing down into a more arenaceous bed. This is separated by a thin layer of clay from another sandy bed containing few fossils		6
(11)	Bluish clay	1	0
(12)	Sandy shales with few fossils	1	0
(13)	Dark-blue clays	9	0
(14)	Fissile sandy shales with *Astarte supracorallina* in abundance		9
(15)	Dark clay	3	0
		43	0

The clays with *Discina* (1) occur only in the western part of the section, and the septarian nodules (2) also disappear near the middle portion of the section, on the north side of the pit. The lowermost **12** feet of the above section is seen in an excavation which has recently been made in the floor near the north-west corner of the pit. The following fossils have been collected from the above beds:—

From the papery shale (1):—

> *Cardium striatulum,* Sow.
> *Lucina minuscula,* Blake
> ,, sp.
> *Arca rhomboidalis,* Contej.
> *Discina latissima* (Sow.)

The upper septarian bed (2). The nodules themselves rarely contain fossils: the following come from the calcareous clay between the nodules:—

> *Ammonites biplex,* Sow.
> ,, *longispinus,* Sow.
> *Cerithium multiplicatum,* Blake
> *Delphinula nassoides,* Buv.
> *Nucula Menki,* Rœm.
> *Cardium striatulum,* Sow.
> *Arca mosensis,* Buv.
> ,, *rhomboidalis,* Contej.
> ,, *rustica,* Contej.
> *Exogyra nana* (Sow.)
> ,, *virgula* (Defr.)

Clays with *Exogyra virgula* in abundance (3):—

> *Ammonites longispinus,* Sow.
> *Aptychi*
> *Thracia depressa* (Sow.)
> *Nucula Menki,* Rœm.
> *Arca rhomboidalis,* Contej.
> *Cardium striatulum,* Sow.
> *Exogyra virgula* (Defr.)
> *Serpula tetragona,* Sow.

Clays with *Ammonites alternans, etc.* (4) :—

> *Ammonites alternans,* Von Buch
> *Alaria* ?
> *Thracia depressa* (Sow.)
> *Cardium striatulum,* Sow.
> *Pecten demissus,* Phil.
> *Exogyra virgula* (Defr.)
> *Lingula ovalis,* Sow.
> *Serpula tetragona,* Sow.

The ferruginous layer (5) yielded no fossils.

Greyish shales (6 and 7) :—

> *Ammonites alternans,* Von Buch
> *Aptychi* (very common in 7)
> *Thracia depressa* (Sow.)
> *Avicula œdilignensis,* Blake
> *Ostrea læviuscula,* Sow.
> *Lingula ovalis,* Sow.
> *Terebratula Gesneri,* Etal.
> *Serpula tetragona,* Sow.

The second septarian layer (8) contained but few fossils. I found in it some *Aptychi* and *Serpula tetragona.*

The thin shaly and arenaceous bed (10), clay (11) and sandy clay (12) contained the following fossils :—

> *Ammonites alternans,* Von Buch
> „   *longispinus,* Sow.
> *Aptychi*
> *Thracia depressa* (Sow.)
> *Cardium striatulum,* Sow.
> *Ostrea læviuscula,* Sow.
> *Rhynchonella pinguis* ?, Rœm.
> *Lingula ovalis,* Sow.

The fossils in the dark-blue clay (13) are fragmentary: in it *Am. alternans* and *Lucina minuscula* were found.

The sandy shales (14) yielded many fossils :—

> *Ammonites calisto,* d'Orb.
> *Thracia depressa* (Sow.)
> *Astarte supracorallina,* d'Orb.

Nucula obliquata, Blake
Arca rhomboidalis, Contej.
Pecten demissus, Phil.
Exogyra virgula (Defr.)
Lingula ovalis, Sow.

It will be seen from the above lists that there are certain well-marked zones in this clay, characterised by the abundance of some fossil; thus:—

(*a*) The clays in the upper portion of the pit with *Discina latissima*, which as far as I have observed does not occur in lower beds. This may be called the *Discina*-zone.

(*b*) The clays (3) are crowded with *Exogyra virgula*, and in no other part of the section do they occur in such abundance. This, therefore, which may be termed the *Virgula*-zone, forms an easily distinguished horizon in this district. *Exogyra virgula*, however, occurs throughout nearly the whole of the Lower Kimeridge.

(*c*) The clays (4 to 13 inclusive) contain *Ammonites alternans* fairly common, and hence may be called the *Alternans*-zone. In it there is a bed (7) which contains numerous *Aptychi*, and this may be regarded as a sub-zone of the *Alternans*-zone.

(*d*) The shales (14) are characterised by the abundance of *Astarte supracorallina*. I have not discovered this fossil in higher beds, and this zone may be called the *Astarte supracorallina*-zone. This is the lowest zone that can be made out in the Roslyn pit.

(*e*) A still lower zone exists in the Haddenham district, which is characterised by the abundance of *Ostrea deltoidea*, and may on that account be termed the *Deltoidea*-zone.

The Kimeridge Clay has, of late, been worked a little to the east of the Roslyn pit, and immediately to the north-west of " u " in " Ouse " on the Survey Map. The clay seen here is of a black colour, and yielded :—

Ammonites calisto, d'Orb.
Astarte supracorallina, d'Orb. (abundant)

Exogyra virgula (Defr.)
Lingula ovalis, Sow.

The fossils indicate that this clay is on about the same horizon as the lowermost clay described above in the Roslyn pit section.

About two miles further north, near Chettisham station, a good section is seen in the Kimeridge Clay, and the beds exposed here are precisely similar to those in the upper portion of the Roslyn Hill section. Its summit is formed of a septarian nodule band, and below it comes 3 feet of clay crowded with *Exogyra virgula*; this is underlain by 5 feet of clay with *Ammonites alternans*, and at its base is a black fissile shale with *Aptychi*. A second septarian band crops out in the floor of the southern part of the section. The following fossils were collected here :—

> *Ammonites alternans*, Von Buch
> *Aptychi*
> *Mactromya rugosa* (Rœm.)
> *Avicula œdilignensis*, Blake
> „   *dorsetensis*, Blake
> „   *nummulina*, Blake
> *Pecten demissus*, Phil.
> *Ostrea lœviuscula*, Sow.
> *Exogyra virgula* (Defr.)
> „   *nana* (Sow.)
> *Lingula ovalis*, Sow.
> *Serpula tetragona*, Sow.
> *Cidaris spinosa*, Ag.

On the road from Ely to Littleport are three brickpits. The most southerly is situated in the low fen-land, two miles south of Littleport. Below the peat there is seen some reassorted clay, underlain by a grey sandy, very compact limestone, about one foot thick; a thin band of clay is exposed below this limestone. The workmen informed me that this clay was 16 feet thick, and that below it a bed of limestone, 6 to 8 inches thick, was met with. The clay is black and somewhat fissile; it contains some ferruginous patches with small crystals of selenite. Fossils are fragmentary : *Exogyra virgula* occurs, but is rare.

The second pit lies at the bottom of the hill about one and a half miles south of Littleport : in it the following section was observed :—

ft. in.

(1) Bluish clay with few fossils 6 0
(2) Compact greyish-white irregularly bedded lime-
    stone with few fossils 9
(3) Greyish clay passing down into black fossili-
    ferous clay 4 0
(4) A layer of dark-coloured compact calcareous
    nodules 6
(5) Black clay . . . . . . 3 ft. seen

Fossil wood and some nodules of pyrites occur in the clay. The following fossils were collected in this pit:—

> *Ammonites trifidus*, Sow.
> *Arca reticulata*, Blake
> *Myoconcha*, sp.
> *Perna Flambarti*, Dollf.
> *Astarte supracorallina*, d'Orb.
> *Pecten Grenieri*, Contej.
> *Lingula ovalis*, Sow.
> *Cidaris*, sp.

Fragments of a chelonian, as well as *Ichthyosaurus* and *Plesiosaurus* vertebræ have been found in this clay.

The beds in this pit dip north at a low angle; it also occupies higher ground than the pit further south, and the beds exposed in this pit are probably above those of the latter.

The third pit is situated at the fourth milestone from Ely, and is near the summit of the high ground on which the village of Littleport stands. The section seen is as follows:—

ft. in.

(*a*) Soil and gravel
(1) Dark-blue clays with small crystals of selenite
    and few fossils 6 0
(2) Layer of isolated septarian nodules lying in
    greyish sandy and calcareous shale with *Astarte
    supracorallina* 8
(3) Dark clays with some thin arenaceous beds . 11 0
(4) Grey very argillaceous limestone with *Trigonia, etc.* 1 0
(5) Black clays 4 0
(6) Very compact limestone; somewhat nodular in
    one part of the section 1 6
(7) Dark-blue tenacious clays . . . . 3 ft. seen

The beds dip north-west at about 5°.

The following fossils were obtained:—

    (i)   From the dark clays (3):—

> *Ammonites calisto*, d'Orb.
> *Astarte supracorallina*, d'Orb. (abundant)
> *Arca reticulata*, Blake
> *Ostrea læviuscula*, Sow.
> *Lingula ovalis*, Sow.

    (ii)   From the clays in the lower part of the pit:—

> *Pliosaurus brachydeirus*, Owen
> *Belemnites abbreviatus*, Mill.
> *Ammonites trifidus*, Sow.
> *Alaria trifida* (Phil.)
> *Trigonia elongata*, Sow.
> *Thracia depressa* (Sow.)
> *Arca reticulata*, Blake
> *Nucula Menki*, Rœm.
> *Lucina minuscula*, Blake
> *Astarte supracorallina*, d'Orb.
> *Myoconcha*, sp.
> *Pecten Grenieri*, Contej.
> *Exogyra virgula* (Defr.), rare
>   ,,   *nana* (Sow.)
> *Lingula ovalis*, Sow.
> *Rhynchonella inconstans* (Sow.)

The following fossils in the possession of Mr Dennis, the owner of this pit, were also obtained here:—

> *Ichthyosaurus*, sp.
> *Plesiosaurus*, sp.
> *Ammonites mutabilis*, Sow.
>   ,,   *biplex*, Sow.
> *Lima pectiniformis* (Schloth.)
> *Serpula intestinalis*, Phil.

The beds in this pit dip north-west, and are situated on higher ground than those of the second pit described above; they would therefore belong to a higher horizon in the series than those of the latter.

It has been stated that the beds in the second pit dip north,

and that they overlie the beds of the first pit; the following
section, which is a summary of the clays exposed in the three pits,
would represent, in descending order, the whole of the Kimeridge
Clay exposed near Littleport:—

			ft.	in.
	(1)	Black clay	6	0
	(2)	Septarian nodule bed		8
	(3)	Clays with *Astarte supracorallina*, etc. .	11	0
Pit No. 3	(4)	Argillaceous limestone	1	0
	(5)	Black clays	4	0
	(6)	Grey limestone	1	6
	(7)	Black clay	3 ft. seen	
	(8)	Bluish clays	6	0
	(9)	Grey irregularly bedded limestone .		9
Pit No. 2	(10)	Grey and black clays	4	0
	(11)	Calcareous nodules		6
	(12)	Black clay	3 ft. seen	
	(13)	Grey compact limestone . . .	1	0
Pit No. 1	(14)	Black clay	16	0
	(15)	Limestone		6–8
			59	0

It is just possible that the clays (7) and (8) of the above section
may be the same; there is very little doubt, however, about the
position of the others.

All the fossils from the clays of the Littleport pits are Lower
Kimeridge forms, so that these clays must belong to that division
of the Kimeridge. *Ammonites alternans* has not been found at
Littleport: this fossil is fairly common in the *Alternans*-zone at
Ely, hence it must be considered that this zone is absent at Littleport. The underlying zone at Ely is characterised by the abundance of *Astarte supracorallina*. Now this fossil is very common in
the upper part of the Littleport clays, and *Ammonites calisto* is
associated with it, and the same may be said of the last-mentioned
zone at Ely. It seems highly probable therefore that the upper
portion of the clays at Littleport is approximately on the same
horizon as the clays in the lower part of the Roslyn pit at Ely.

The western flank of the ridge of Corallian limestone near
Upware is partly formed of Kimeridge Clay, as stated above
(p. 52). The clay is said to be of a black colour, and at its junction
with the Coral Rag has mixed with it a quantity of broken and

often rounded fragments of Coral Rag. No fossils have been recorded from this clay.

The thickness of the Kimeridge Clay may be approximately estimated from the sections given above. In the following list I have arranged the beds in the order which I consider them to occur:—

		ft.	in.
(a)	The clays at Roslyn pit belong to the highest zone of the Kimeridge Clay exposed in this neighbourhood	43	0
(b)	The Littleport clays probably underlie those of Ely —their thickness is given on p. 73 . . .	59	0
(c)	The clays near Haddenham Station may be lower than those of Littleport	19	9
(d)	The clays, with the phosphatic nodule bed at their base, west of Haddenham, are the lowest beds of the Kimeridge Clay in this district and are lower in the series than those at Haddenham Station .	20	6
		142	3

This must be at least the thickness of the Kimeridge Clay developed in the district under consideration. In the well-boring at Chettering, only 51 feet of the Kimeridge Clay was pierced, and this was probably but little more than that seen in the pits near Haddenham.

Professor Blake places[1] the clays of the lower part of the Roslyn pit (those with *Ammonites alternans*) in the Lower Kimeridge "partly on account of its more clayey character, but chiefly from its contained fossils." The clay with *Exogyra virgula* overlies them, and he states that though this fossil is met with at intervals in the Lower Kimeridge Clay of Dorsetshire, it "is nowhere so plentiful as it is here, where it might justify the title Virgulian for the beds; and we may thus regard its abundance as marking the passage from Lower to Upper Kimeridge[2]."

The papery shale with *Discina latissima*, Professor Blake includes in the Upper Kimeridge, and these form the highest beds of the Kimeridge Clay exposed in this district. It is evident therefore that almost the whole of the Kimeridge Clay described above belongs to the lower division of that formation, certainly those of Haddenham, Littleport and the lower 27 feet of the beds in the Roslyn pit are Lower Kimeridge.

---

[1] *Quart. Journ. Geol. Soc.*, vol. xxxi. (1875), p. 211.　　　[2] *Ibid.* p. 201.

Subjoined is a list of the Kimeridge Clay fossils in the Woodwardian Museum, and also those in the possession of Mr Marshall Fisher at Ely, the latter marked F :—

LIST OF KIMERIDGE CLAY FOSSILS IN THE WOODWARDIAN MUSEUM AND IN MR FISHER'S COLLECTION AT ELY.

	Ely	Chettis-ham	Hadden-ham	Cotten-ham	Stret-ham	Little-port	
REPTILIA.							
Iguanodon, sp.	F						
Gigantosaurus megalonyx, Seeley	×			×	×		
Cimoliosaurus trochanterius (Owen)	×		×				
Plesiosaurus sterrodeirus, Seeley	×						
" sp.	×			×		×	
Pliosaurus brachydeirus, Owen	F			×		×	
" brachyspondylus, Seeley	×						
" grandis, Owen.	F						
" sp.	×		×			F	
Ichthyosaurus chalarodeirus, Seeley		×					
" hygrodeirus, Seeley		×			×		
" sp.	×	×	×		×	×	
Dacosaurus maximus (Plieninger)	×			×			
" sp.	F						
Steneosaurus, sp.	F						
Enaliochelys chelonia, Seeley	×						
Trionyx, sp.						×	
PISCES.							
Asteracanthus ornatissimus, Ag.	×			×			
" carinatus, M'Coy	×						
Ditaxiodus impar, Owen	×						
Ischyodus, sp.	×						
Gyrodus umbilicatus, Ag.	×						
Hybodus acutus, Ag.	F						
Lepidotus, sp.	×						
Pachycormus, sp.	. ×						
Eurycormus grandis, Woodw.	×						
Leptacanthus semicostatus, M'Coy	×						
Sphenonchus, sp.	F						
CEPHALOPODA.							
Belemnites abbreviatus, Mill.	×		×			×	
" explanatus ?, Phil.	×						
" nitidus, Dollf.			×				
Ammonites biplex, Sow.	×		×			×	
" trifidus, Sow.						×	
" mutabilis, Sow.		×	×			×	
" longispinus, Sow.	×	×					
" rotundus, Sow.	×						
" cordatus, Sow.	×						
" alternans, Von Buch	×	×					
" calisto, d'Orb.	×					×	
" eudoxus, d'Orb.	×						
Trigonellites (Aptychus) latus, Park.	×	×					

	Ely	Chettisham	Haddenham	Cottenham	Stretham	Littleport
GASTEROPODA.						
Alaria trifida (Phil.)	x		x			x
Delphinula nassoides, Buv.	x					
LAMELLIBRANCHIATA.						
Cardium striatulum, Sow.	x					
Lucina minuscula, Blake	x					x
Astarte supracorallina, d'Orb.	x		x			x
Myoconcha, sp.						x
Pleuromya donacina, Ag.			x			
Mactromya rugosa (Rœm.)		x				
Pholadomya ovalis, Sow.			x			
Thracia depressa (Sow.)	x		x			x
Myacites, sp.			x			
Trigonia elongata, Sow.	x		x			x
„ *Pellati,* Mun. Chal.	x		x			
Nucula Menki, Rœm.	x		x		x	x
„ *obliquata,* Blake	x					
Arca mosensis, Buv.	x		x			
„ *rustica,* Contej.	x					
„ *rhomboidalis,* Contej.	x		x			
„ *reticulata,* Blake						x
Avicula œdilignensis, Blake	x	x	x			
„ *nummulina,* Blake	x	x				
„ *dorsetensis,* Blake	x	x	x			
Perna Flambarti, Dollf.						x
Lima pectiniformis (Schloth.)						x
Pecten Grenieri, Contej.			x			x
„ *lens,* Sow.	x		x		x	
„ *demissus,* Phil.	x	x				
„ sp.	x					
Anomia Dollfussi, Blake			x			
Exogyra virgula (Defr.)	x	x	x			x
„ *nana* (Sow.)	x	x	x			x
Gryphœa dilatata ?, Sow.	x					
Ostrea deltoidea, Sow.			x			
„ *gregaria,* Sow.	x		x			
„ *monsbeliardensis,* Contej.			x			
„ *solitaria,* Sow.			x			
„ *lœviuscula,* Sow.	x	x	x			x
BRACHIOPODA.						
Terebratula Gesneri, Etal.	x		x			
Rhynchonella inconstans (Sow.)	x	x				
„ *pinguis*? (Rœmer)	x					
Discina latissima (Sow.)	x					
„ small sp.			x			
Lingula ovalis, Sow.	x	x				x
VERMES.						
Serpula tetragona, Sow.	x	x	x			
„ *intestinalis,* Phil.			x			x
Vermicularia contorta, Blake	x					
ECHINOIDEA.						
Cidaris spinosa, Ag.	x	x				
„ sp.			x			x
CIRRIPEDIA.						
Pollicipes Hausmanni, Koch & Dun.	x		x			

VIII. CORRELATION WITH OTHER ENGLISH DEPOSITS.

THE upper and lower members of the Jurassic series represented in the Cambridgeshire district do not differ very much
from their equivalents in other parts of England: it will be seen,
however, that the intermediate groups are peculiar and differ very
markedly from beds of the same age in other areas.

Oxford Clay.—This, as already stated, is fairly constant in
character wherever it has been met with in England. In Dorsetshire the clay itself and its fossils are much the same as those of
Huntingdonshire and Cambridgeshire, and the same may be said
of these clays all through Wiltshire, Oxfordshire, Bedfordshire and
Lincolnshire. Professor Phillips[1] divides the Oxford Clay near
Oxford into :—

Upper part with *Ammonites vertebralis*
Middle   „   „   *Lamberti*
Lower   „   „   *Duncani*

and this classification would apply equally well for the Cambridgeshire district.

In Yorkshire, the Oxford Clay is subject to some variations.
The calcareous and arenaceous division at its base (the Kellaways Rock) sometimes attains a thickness of 50 feet, as at Scar-
borough; it thins out in a south-easterly direction, and at Gris-
thorpe Bay it is only some 5 feet thick[2]. Thus, in part of that
district the lower portion of the Oxford Clay series may either be
composed almost entirely of shales, or it may be largely composed
of sandstone and grits. The upper part of the Oxford Clay is
sometimes similarly replaced by the Lower Calcareous Grit[3]. In

[1] *Geology of Oxford and the Valley of the Thames,* (1871), p. 298.
[2] W. H. Hudleston, *Proc. Geol. Assoc.,* vol. IV. (1875), p. 354.
[3] *Ibid.* p. 383.

that county too, the Oxford Clay itself is much more sandy than it
is elsewhere in England.

Mr Hudleston gives[1] the following classification of the Oxfordian
series in Yorkshire :—

Formation	*Some of the Characteristic Fossils*
Lower Calcareous Grit	

Oxford Clay
{ Upper—*Am. perarmatus*, Sow., rarely.
Middle—Small Ammonites, young of *Am.
Eugenii*, Rasp., *A. crenatus*, Brug., etc.
Lower—*Bel. Oweni*, Pratt, *A. Lamberti*, Sow.,
A. athletus, Phil., *A. oculatus*, Phil., *A.
crenatus*, Brug.

Kellaways Rock
{ *Bel. Oweni*, var. *tornatilis*, Phil., *Am. Jason*,
Rein., *A. Duncani*, Sow., and var. *gemi-
natus*, Phil., *A. athletus*, Phil., *A. Hecticus*,
Rein., *A. lunula*, Rein., *A. Gowerianus*,
Sow., *A. Kœnigi*, Sow., etc.

It will be seen from a comparison with the fossil list given on
pages 17 and 18 that all the characteristic fossils mentioned above
from the so-called Oxford Clay of Yorkshire are found in the clays
at St Ives, that is to say, in the upper part of the Oxford Clay of
Huntingdonshire. Again, some of the Ammonites (*A. Jason,
Duncani, athletus*) recorded from the Kellaways Rock occur low
down in the Oxford Clay of this district (*e.g.* at St Neots). It
would appear, therefore, that the clays at St Neots are on the
same horizon as a portion of the Kellaways Rock of Yorkshire, and
that the ' Oxford Clay ' of Yorkshire represents only a part of the
clays of that formation in the neighbourhood of Cambridge[2].

In the Cave district (South Yorkshire) the Oxford Clay is
described as resembling that of St Ives, rather than that of north-
east Yorkshire[3].

Lower Calcareous Grit.—The position of the Elsworth and
St Ives Rocks has already been discussed, and the conclusion
arrived at was that they represent, in part at least, the Lower

[1] *Geol. Mag.*, dec. 2, vol. ix. (1882), p. 147.
[2] *Proc. Geol. Assoc.*, vol. iv. (1875), p. 410.
[3] W. Keeping and C. S. Middlemiss, *Geol. Mag.*, dec. 2, vol. x. (1883), p. 218;
A. Harker, *Naturalist*, (1885), p. 231.

Calcareous Grit. If this be their true position, these limestones but feebly represent the rocks of this age in other areas.

In the Weymouth district, the Lower Calcareous Grit is represented by :—

	feet
Bengliff Grits	21
Nothe Clays	40
Nothe Grits	30

Several of the fossils in the last two subdivisions occur in the Elsworth and St Ives Rocks. Out of the 21 species recorded[1] from the Nothe Grits, 12 occur in these limestones, and 10 out of the 16 species in the Nothe Clays.

In the southern part of North Dorsetshire the Lower Calcareous Grits are but feebly represented, but proceeding northwards they assume more importance. In Oxfordshire these beds are composed of " sands partly consolidated into sandstones," and sometimes they are calcareous. Near Oxford they are about 60 to 70 feet thick, and can be traced as far as Wheatley (north-east Oxfordshire), and there the Corallian Rocks " die out, not gradually but suddenly, and the normal pelolithic formation reigns supreme[2]."

The hard dark-blue limestone at the base of the Corallian in the section at Stanton St John, and a similar limestone at Studley, may be the equivalent of the Elsworth Rock. The following fossils are recorded[3] from Studley :—*Ammonites vertebralis, Pholadomya, Pinna, etc.*

Beyond this, except the limestones in question of the Cambridgeshire district, the Lower Calcareous Grits have not been observed until Yorkshire is reached, and there they are well developed, being represented by calcareous grits and sands from 80 to 100 feet thick. Neither in this district, nor in the south of England, has any bed possessing the characters of the Elsworth and St Ives Rocks been described of Lower Calcareous Grit age.

In the sub-Wealden boring the Lower Calcareous Grit was either absent or was represented by sandy calcareous shales[4]. A

---

[1] *Quart. Journ. Geol. Soc.*, vol. xxxiii. (1877), p. 263.   [2] *Ibid.* p. 311.

[3] J. Phillips, *Geology of Oxford and the Valley of the Thames*, (1871), p. 298; A. H. Green, *Geology of the country around Banbury, Woodstock, Bicester, and Buckingham*, (1864), p. 44.

[4] W. Topley, *Rep. Brit. Assoc.*, 1875 (1876), p. 348,

band of oolitic limestone was pierced, immediately before the dark shales, which are considered to be Oxfordian, were reached. This oolitic limestone may be the equivalent of the Elsworth Rock; no fossils, however, are recorded from it, and the only evidence for this correlation is the superposition of the beds.

Ampthill Clay.—It has been shown that this clay overlies the representative, in part at least, of the Lower Calcareous Grit, and is succeeded by the basement bed of the Kimeridge Clay. From its stratigraphical position it must, therefore, be regarded as Corallian in age. The evidence derived from the fossils also support this view.

The Corallian limestones, clays and grits, so well developed in the southern counties of England and Yorkshire, are absent as such also in the counties of Lincoln, Bedford, Buckingham and part of Oxford, as well as in the Wealden area. There is reason for believing, however, that in some of the districts just mentioned a series of clays is present between the Oxford and Kimeridge Clays, and which are identical in character with the Ampthill Clays of Cambridgeshire.

In Lincolnshire[1] there is a band of black seleniferous clays overlying the uppermost zone of the Oxford Clay with *Cordati* Ammonites. These black clays are worked in the brickyards of Hawkstead Hall, south-east of Bardney, North and South Kelsey, and are exposed in the railway cutting east of Brigg, and in the two cuttings west of Wrawby Road Bridge. They were also met with in a well-boring between Market Rasen and Bishop's Bridge. They are precisely similar lithologically to the Ampthill Clays, and contain such fossils as *Ammonites plicatilis, Ostrea deltoidea,* and *Gryphœa dilatata* and other species, nearly all of which occur in the Ampthill Clay, and also in the typical Corallian rocks of the southern counties. They are succeeded by clays containing *Ostrea deltoidea* in abundance, and of the same character as the lowermost beds of the Kimeridge Clay of Cambridgeshire.

Tracing the Ampthill Clay south of the Cambridgeshire district, the first locality where it is known to occur is in the railway cuttings on each side of the tunnel at Ampthill Park, Bedfordshire. In the northern cutting are some brownish clays with selenite crystals in the upper part of the section; below come dark-blue

---

[1] T. Roberts, *Quart. Journ. Geol. Soc.*, vol. XLV. (1889), p. 545.

clays with several (10) bands of limestone varying from 6 inches to 1 foot in thickness; some of the limestones are nodular. The beds dip south-east at a low angle. In the southern cutting black clays only are seen. The following fossils were collected from this locality by Mr A. C. Seward, and presented to the Woodwardian Museum :—

> *Belemnites hastatus* (Montf.)
> *Ammonites alternans*, Von Buch
>   „   *vertebralis*, Sow.
>   „   *excavatus*, Sow.
> *Alaria trifida* (Phil.)
> *Cucullœa concinna*, Phil.
> *Arca*, sp.
> *Avicula œdilignensis*, Blake
> *Nucula Menki*, Rœm.
> *Thracia depressa* (Sow.)
> *Trigonia* (fragment of a clavellate species)
> *Pecten articulatus* (Schloth.)
> *Ostrea gregaria*, Sow.
>   „   sp.
> *Gryphœa dilatata*, Sow.
> *Rhynchonella*, sp.
> *Serpula tetragona*, Sow.

Further south these clays do not appear to have been recognised with certainty.

In the Survey Memoir[1] on sheet 45, it is stated that at Studley is " a peculiar bed of clouded grey colour, and very tough and dense texture, a sort of argillaceous chert, rich in *Pinnœ*, *Ammonites*, and other organic remains." (See p. 79). " This bed runs with a good escarpment by Arngrove Farm to the north of the Gravel Pit Farm, beyond which point we lose sight of it altogether. It dips gently to the east, but whether it runs under the clay beds to the east of it, or passes into clay, it is not easy to say. Immediately below this stone we find Oxford Clay with *Gryphœa dilatata*, and it is therefore without doubt the bottom bed of the Calcareous Grit."

" To the east of this is a band of very marked light-blue clay,

---

[1] A. H. Green, *Geology of the country around Banbury, Woodstock, Bicester, and Buckingham* (1864), p. 44.

82   CORRELATION WITH OTHER ENGLISH DEPOSITS.

somewhat sandy, and in places crowded with *Ostrea sandalina*. This bed ranges through Worminghall, Oakley, and Boarstall, and was again found with its characteristic fossil at Westcot, and one mile west of Waddesdon Field. On the strength of this evidence a belt of this clay has been drawn between the Kimeridge and Oxford Clays up to the last-named spot, from whence it has been supposed to thin away towards the north-east, and it has been looked upon as the representative of the Coralline Oolite, (1) from its position, lying as it does between the Oxford and Kimeridge Clays, and differing in mineral character from both ; (2) from the abundance of *Ostrea sandalina* found in it, that shell being plentiful in the lower part of the Coral Rag." The following fossils are given (p. 45) from these localities :—

> *Ammonites Lamberti*, Sow.
> ×   „   *cordatus*, Sow.
> × *Belemnites abbreviatus*, Mill.
> × *Gryphœa dilatata*, Sow.
> × *Ostrea gregaria*, Sow.
>   „   *sandalina*, Goldf.
>   „   sp.
> *Serpula*, sp.
> × *Pentacrinus*, sp.

It is stated[1] that this clay may be the equivalent of the Tetworth (Ampthill) Clay, with which view I agree. Nevertheless they differ in their lithological character, and to some extent in their fauna. The fossils marked × occur in the Ampthill Clay; a small oyster, which may be the same as that referred to *O. sandalina*, occurs in this clay at Ampthill in abundance, but is not well preserved. This is rather a high horizon for *Ammonites Lamberti*, since it has, as yet, only been recorded from the Oxford Clay.

In the Weymouth district there is a local development of clay representing a portion of the Corallian Series. This clay, which occurs more especially at Sandsfoot Castle, is sometimes 40 feet thick, but is much thinner at Wyke, Linton Hill and Osmington[2]. The fossils recorded from it are dwarfed deltoid

[1] A. H. Green, *Geology of the country around Banbury, Woodstock, Bicester, and Buckingham* (1864), p. 45.
[2] *Quart. Journ. Geol. Soc.*, vol. xxxiii. (1877), p. 269.

oysters, *Astarte supracorallina, Corbula Deshayesea, Nucula Menki,* a small *Arca* and a *Trigonia.* The clays are underlain by the *Trigonia*-beds which Messrs Blake and Hudleston[1] regard as representing the upper portion of the Coralline Oolite; the overlying beds are the Sandsfoot Grits, which though containing some Kimeridgian species such as *Astarte supracorallina, Ostrea deltoidea, Discina Humphriesiana* and *Lingula ovalis,* nevertheless have yielded some true Coral Rag forms such as *Cidaris Smithi* and *Cidaris florigemma* and must be of Upper Corallian age. The clays of Sandsfoot Castle must, therefore, lie in the midst of the Corallian Series. The fossils which occur in these clays are also found in the Ampthill Clay and on this account as well as from their position, the Sandsfoot clays must be regarded as a local development of those of Ampthill.

In the sub-Wealden boring, all the beds pierced between 274 feet from the surface and the greatest depth reached (1870 feet) are considered to be the representatives of the Kimeridge Clay, the Corallian and part of the Oxford Clay, though a good deal of doubt exists as to where the lines should be drawn separating these formations. Mr Topley[2] states that at 1769 feet from the surface, an oolitic rock was reached which proved to be 17 feet thick; this he considers to be Coralline Oolite. It was not very fossiliferous and in the boring contained only small oysters. Beds with oolitic structure also occurred lower down, and these may be a thin representative of the Lower Calcareous Grit; in the absence of fossils, however, it is impossible to assign them to their proper horizon. The Clay at the bottom of the boring is considered to be Oxfordian.

If the limestone that was met with at 1769 feet be Coralline Oolite, the clays, with some limestones and sandstones, which come between it and the base of the Portland, that is to say, between 275 and 1769 feet from the surface, may represent the rest of the Corallian, together with the whole of the Kimeridgian, or the former may be absent and the whole would then be Kimeridgian.

In Dixon's 'Geology of Sussex[3],' it is stated on the authority

---

[1] *Quart. Journ. Geol. Soc.,* vol. xxxiii. (1877), p. 389.

[2] W. Topley, *Rep. Brit. Assoc.* 1874 (1875), p. 348.

[3] F. Dixon, *Geology of Sussex* (1878), p. 155.

of Mr Hudleston, that some of the sandstone beds at from 1070
feet from the top of the Kimeridge Clay (1345 feet from the
surface) may be Corallian.

The following fossils, however, are recorded from beds below
these sandstones [1]:

Exogyra virgula (Defr.)
Cardium striatulum, Sow. } at 1634 ft. from the surface

Rhynchonella pinguis (Rœmer)
Ostrea bruntrutana (Thurm.) } from 1656 ft. below the surface.
Pinna lanceolata, Sow.

All these species are characteristic of the Lower Kimeridge
Clay, and the beds to this depth must be included in that formation. Only one species (*Trichites Plotti*) is given from a lower
depth (1661 feet), except some Oxford Clay forms from near
the bottom of the boring.

From these considerations, it would appear, that all the beds
above the so-called Coralline Oolite up to the base of the Portlandian, are Kimeridge Clay, and that there is no representative
of the Ampthill Clay, unless indeed the shaly beds below the
'Coralline Oolite' be considered as such; no fossils are recorded
from these shales and the only evidence for placing them in
this horizon is obtained from the position of the beds, occurring,
as they are supposed to do, between the 'Coralline Oolite' and
Oxford Clay.

The Corallian Rocks of Upware.—The correlation of these
beds has been given by Messrs Blake and Hudleston [2]. The
limestones of the south quarry are placed in the true Coral Rag
with *Cidaris florigemma,* and those of the north quarry in the
underlying Coralline Oolite. This arrangement of the beds, as
is pointed out by them, is found in Yorkshire, and again in the
south of England. The Coralline Oolite, however, in these
localities, is somewhat variable being at times represented by
a massive limestone and at others by a marl [3]. Similarly the
Coral Rag does not always present the same character as at
Upware.

---

[1] W. Topley, *Rep. Brit. Assoc.* 1874 (1875), p. 26.
[2] *Quart. Journ. Geol. Soc.,* vol. xxxiii. (1877), p. 313 *et seq.*
[3] *Ibid.* p. 389.

It is worthy of notice that two species of *Hyboclypus* (*H. gibberulus* and a new species) occur at Upware, which have not been recorded from these beds elsewhere in England.

Kimeridge Clay.—Professor Blake divides[1] the Kimeridge Clay of England into an upper and a lower division, and, in districts where it is preceded by the Coral Rag, he adds a series of beds which he calls the Kimeridge Passage-beds.

Blake's classification of the beds in the Roslyn pit has already been given. The *Discina*-zone is placed in the Upper Kimeridge; the *Virgula*-zone, he considers, marks the passage from the Lower to the Upper Kimeridge; whilst the clays below are included in the Lower Kimeridge. The clays at Littleport and Haddenham come below those of Ely, and, as already stated, must be placed in the Lower Kimeridge.

Exogyra virgula occurs nearly throughout the whole of the Lower Kimeridge of this district, but I have not met with it in the *Discina*-zone; the lithological character of the *Virgula*-zone at Ely, too, is much more closely allied to the underlying than to the overlying beds, and the more natural classification for this district would be, to draw the line between the Lower and Upper Kimeridge at the base of the *Discina*-zone, and include the *Virgula*-zone in the Lower Kimeridge. In Dorsetshire *Exogyra virgula* occurs throughout the Lower Kimeridge Clay and has been found only in the lower portion of the Upper Kimeridge; it seems, therefore, to be more characteristic of the lower than of the upper division of this formation. If this classification be accepted, the *Discina*-zone will be the only representative of the Upper Kimeridge in this district.

The basement bed of the Kimeridge Clay of Cambridgeshire is very similar to that which occurs further south. Near Oxford, according to Professor Phillips[2], it contains phosphatic nodules. At the Headington pits, near Oxford, the Kimeridge Clay is seen resting on the Coralline Oolite. "About eight feet above the junction is a calcareous band eight inches thick...... There are probably two or three beds of *Ostrea deltoidea*, one near the base being often recognised, even as far north as Yorkshire[3]."

[1] *Quart. Journ. Geol. Soc.*, vol. XXXI. (1875), p. 196.   [2] *Ibid.* vol. XIV. (1858), p. 236.
[3] J. Phillips, *Geology of Oxford and the Valley of the Thames* (1871), p. 325.

On the Dorsetshire coast, *Ostrea deltoidea* is found[1] chiefly at the base of the Lower Kimeridge. In the succeeding clays the characteristic fossil[2] appears to be *Ammonites alternans*, whilst in the overlying clays this fossil is absent and *Exogyra virgula* is fairly common[3]. Following on these comes the Upper Kimeridge Clay with *Discina latissima*. In this area therefore the upper and lower zones of the Lower Kimeridge Clay are similar to those of Cambridge but the fossils have not the same distribution in the intermediate zones.

Overlying the seleniferous clays of Lincolnshire, there is, as already stated, a zone of clays crowded with *Ostrea deltoidea* and these are succeeded by clays in which *Ammonites alternans* is fairly common: these clays are exposed in the pits of Baumber, Hatton, and Market Rasen. The last named fossil is absent in the next zone of clays which are seen in the brickyards near Horncastle, and this is the highest zone as yet recognised in the Lower Kimeridge of Lincolnshire. The Upper Kimeridge in this county is similar to that of the southern counties and contains *Discina latissima* in abundance. It is well seen in the clay-pit west of Fulletby. The zones of the Lower Kimeridge Clay of Lincolnshire agree more closely with those of Dorsetshire than with those of Cambridgeshire.

In the sub-Wealden boring a bed with *Exogyra virgula* was met with at a depth of 1126 feet from the surface. This occurs probably at a lower horizon than the *Virgula*-zone at Ely, because some true Lower Kimeridge fossils were found in the clays above it. Amongst others *Astarte supracorallina* is recorded from 926 feet from the surface, and this fossil has not been found elsewhere in England in beds higher than the Lower Kimeridge.

From the foregoing considerations, it will be seen that the fossiliferous zones in the Lower Kimeridge are not constant over wide areas and can only be used locally[4].

---

[1] *Quart. Journ. Geol. Soc.*, vol. XXXI. (1875), p. 212.

[2] W. Waagen, *Versuch einer allgemeinen Classification der Schichten des oberen Jura* (1865), p. 5.   [3] *Ibid.* p. 6.

[4] See also J. F. Blake, *Quart. Journ. Geol. Soc.*, vol. XXXI. (1875), p. 197.

IX. Correlation with the Foreign Deposits.

(a) *The north and north-west of France or the Paris Basin.*

The correlation of the Upper Jurassic Rocks of the Paris basin with their English equivalents, has been worked out by Professor Blake[1]. The following are the larger divisions in that district:—

1. Purbeckian
2. Portlandian
3. Kimeridgian
4. Corallian
5. Oxfordian

These divisions do not correspond exactly with those which are similarly named in England; as for instance a portion of the Corallian of England is included in the French Oxfordian, and their Portlandian includes some part of our Kimeridgian.

To prevent any further confusion Professor Blake proposes the following classification[2] for these beds:

1. Portlandian $\begin{cases} \text{Upper} = \text{Purbeck} \\ \text{Lower} = \text{Portland Limestone} \end{cases}$

2. Bolonian $\begin{cases} \text{Upper} = \text{Middle Portland} \\ \text{Lower} = \text{Lower Portland} \end{cases}$

---

[1] *Quart. Journ. Geol. Soc.*, vol. xxxvii. (1881), p. 497.
[2] *Ibid.* p. 567.

3. Kimeridgian $\begin{cases} \text{Virgulian} \\ \text{Pterocerian} \\ \text{Astartian} \end{cases}$

4. Corallian $\begin{cases} \text{Supracoralline} \\ \text{Coral Rag} \\ \text{Coralline Oolite} \end{cases}$

5. Oxfordian $\begin{cases} \text{Upper} = \begin{cases} \text{Oxford Oolite} \\ \text{Oxford Grit} \end{cases} {}_1 \\ \text{Lower} = \text{Oxford Clay} \end{cases}$

The Lower Oxfordian is considered to be the equivalent of the Oxford Clay of England. As a rule it is composed of clay with a calcareous deposit (the Callovian) at its base, so that the lithological resemblance of this deposit in France and England is great. The fossils, too, are similar.

In the Upper Oxfordian is included the Lower Calcareous Grit and part of the overlying limestone. The usual type of the Upper Oxfordian of France is a richly fossiliferous ironstone. Near St Mihiel it is a ferruginous oolite crammed with fossils. Amongst others the following are recorded :—

> × *Ammonites convolutus?*, Quenst.
> *Pholadomya decemcostata*, Rœm.
>   ,,   *deltoidea* (Sow.)
> *Mytilus pectinatus*, Sow.
> *Perna quadrata*, Sow.
> × *Pecten articulatus* (Schloth.)
> × *Ostrea dilatata* (Sow.)
> *Rhynchonella lacunosa* (Schloth.)
> × *Terebratula bucculenta*, Sow.
> × *Collyrites bicordata* (Leske)
> *Dysaster ovalis* (Leske)

This limestone resembles, to some extent, those of Elsworth and St Ives ; five of the above fossils (those marked ×) are common to the deposits in the two areas.

Dr Davidson states that[2] "during a recent visit to Cambridge Mr E. Rigaux examined the Elsworth Rock and recognised it as

---

[1] These include the Lower Calcareous Grit and some part of the overlying limestone.

[2] *Mon. Brit. Foss. Brach.*, vol. IV. (1878), p. 173.

the equivalent of his *Calcaire d'Houllefort*, the stratigraphical position of which is immediately above the Oxford Clay bed containing *Waldheimia impressa* and *Terebratula insignis.*" Professor Blake[1], referring to this limestone of Houllefort, considers that, although its fauna is somewhat peculiar, "the facies is Corallian, and indicates an horizon which may indeed be compared with that of Neuvizy or the Osmington Oolite, but is more like the base of the Corallian in the Haute-Marne and the Yonne departments."

The Corallian Series, though very varied in its composition, is for the most part formed of limestones; there is, therefore, no formation on this horizon in France, which shews any resemblance to the Ampthill Clay. The Coral Rag and Coralline Oolite of Upware, Blake includes in the middle and lower divisions of his Corallian.

The English equivalents of the Astartian beds of France are, by Professor Blake, stated to be the Kimeridge Passage-beds, together with the lower portion of the Kimeridge Clay containing *Ostrea deltoidea* and *Rhynchonella inconstans.* Since there are no Kimeridge Passage-beds in the Cambridgeshire district, the *Deltoidea*-zone would be the sole representative of the Astartian.

The whole of the Lower Kimeridge was formerly included in the Astartian on account of the occurrence in it of an Astarte, which is like *A. supracorallina*, though it may be distinct[2].

All the Lower Kimeridge above the *Deltoidea*-zone, Professor Blake correlates with the Pterocerian and Virgulian. The subdivision of the English beds into these two stages, however, is not very clear: he states[3] that "it may, perhaps, fairly be taken that the absence of *Exogyra virgula* (as in Lincolnshire) indicates the former [Pterocerian], whilst its abundance (as at Ely, Swindon, and in parts of the southern coast section) indicates the latter [Virgulian]."

It has been usually considered by English geologists that there is no representative of the Pterocerian in this country; and no true Pterocerian fauna has been found in England on the horizon claimed for it by Professor Blake. There is certainly none in the

[1] *Quart. Journ. Geol. Soc.*, vol. xxxvii. (1881), p. 560.
[2] J. F. Blake, *Ibid.* p. 580.
[3] *Ibid.*

Cambridgeshire district, since *Exogyra virgula* occurs throughout the Kimeridge Clay of this district, except in the basement beds with *Ostrea deltoidea;* consequently the whole of it would be Virgulian, if *Exogyra virgula* be considered characteristic of and peculiar to that stage.

The thin representative of the Upper Kimeridge Clay in the upper portion of the section at Roslyn pit would be included in the lowermost division of the Bolonian of Blake. No higher beds of the Jurassic rocks occur in the district under consideration.

(*b*) *The Jura.*

I have elsewhere[1] attempted a correlation of the Upper Jurassic Rocks of the Swiss Jura with those of England, and the equivalents of the Oxford Clay, Corallian and Kimeridge Clay of England were shown to be as follows :—

	ENGLAND.		SWISS JURA.	
	Upper Kimeridge Clay.		Portlandien (in part).	
LOWER KIMERIDGE.	Clays with *Exogyra virgula.*		Virgulien.	
	„ „ *Ammonites alternans.*		Ptérocérien.	
	Clays with *Astarte supracorallina.*			
	„ „ *Ostrea deltoidea.*		Astartien.	
	Kimeridge Passage-beds.			
CORALLIAN.	Supracoralline.		Calcaire à Nérinées.	CORALLIEN.
	Coral Rag.		Oolithe Corallienne.	
	Coralline Oolite.		Terrain à chailles siliceux.	
	Middle Calcareous Grit.			
	Hambleton Oolite.		Pholadomien.	OXFORDIEN.
	Lower Calcareous Grit.		Spongitien.	
OXFORD CLAY.	Clays with *Cordati* Ammonites.		Le fer sous-oxfordien.	CALLOVIEN.
	„ „ *Ornati* Ammonites.		Zone of *Amm. macrocephalus.*	
	Kellaways Rock.			

The Oxford Clay of the Cambridge district is the equivalent of the *Fer sous-oxfordien* and its upper part is probably the repre-

[1] *Quart. Journ. Geol. Soc.,* vol. XLIII. (1887), p. 229.

sentative also of part of the Oxfordian of the Jura, whilst the Elsworth and St Ives Rocks are synchronous with the remaining portion of the Oxfordian.

The Upware Coral Rag and Coralline Oolite are correlated with the *Terrain à chailles siliceux* and part of the overlying *Oolithe Corallienne* and *Calcaire à Nérinées*, to the latter of which our Coral Rag presents several points of affinity, more especially in its lithological character.

The equivalents of the Cambridgeshire Kimeridge Clay are indicated in the above table. The absence of Supracoralline Beds and also the Kimeridge Passage-beds from this district does not make the correlations as complete as that in other parts of England.

(c) *The north-west German area.*

The following classification is given by **Dr C. Struckmann**[1] of the Upper Jurassics of Hanover:—

	Purbeck (Serpulite) or Lower Wealden
Portland	⎰ Münder-Mergel as transition between Purbeck and Portland ⎱ Eimbeckhäuser Plattenkalke ⎰ Beds with *Ammonites gigas*
Kimeridge	⎰ Upper Kimeridge or *Virgula*-beds ⎱ Middle Kimeridge or *Pteroceras*-beds ⎰ Lower Kimeridge (Astartian) *i.e. Nerinœa*-beds and zone of *Terebratula humeralis*
Corallian	⎰ Upper Coralline Oolite (zone of *Pecten varians*) ⎱ Lower Coralline Oolite (zone of *Ostrea rastellaris* and Coral-bed)
Oxford group	Hersumer beds
Kellaway group of the Middle Jurassic	⎰ (a) *Ornatus*-clays ⎱ (b) *Macrocephalus*-beds

The *Macrocephalus*-beds are the equivalents, in part at least, of the Kellaways Rock. The *Ornatus*-clays contain *Ammonites Lamberti, A. Jason, A. ornatus, etc.,* and may be correlated with some of the Lower Oxford Clay in the Cambridgeshire district. Mr Hudleston[2] is of opinion that in Yorkshire the line of division

---

[1] *Neues Jahrbuch für Mineralogie, etc.* Bd. II. (1881), p. 77; *Geol. Mag.*, dec. 2, vol. VIII. (1881), p. 546.

[2] *Geol. Mag.*, dec. 2, vol. IX. (1882), p. 148.

between the Brown (Middle) and White (Upper) Jura of the
Germans, that is to say, the upper limit of the *Ornatus*-clays,
" would probably run somewhere in the great unfossiliferous masses
of the upper part of the 'Oxford Clay.'" It has been shewn that
the so-called 'Oxford Clay' of Yorkshire is the equivalent of the
upper part of that series in this district, *i.e.* the clays at St Ives;
consequently the line of division in question would be drawn in
this district, through some part of the clay in the St Ives pit.
The fossil zones, however, have not been worked in this pit, and
consequently its exact position cannot be given.

Struckmann correlates the Hersumer Beds as follows :—

English Equivalents

Hersumer Beds
(a)
  (ii) The Lower Limestones or Hambleton Oolite
  (i) The Lower Calcareous Grit
(b) Upper part of the Oxford Clay with *Ammonites cordatus, A. athletus,* and *Ostrea dilatata*

In the Cambridgeshire district, according to this view, the
Hersumer Beds would be represented by the upper part of the
Oxford Clay, the Elsworth and St Ives Rocks, and probably some
portion of the Ampthill Clay.

The Lower Coralline Oolite (zone of *Ostrea rastellaris* and
Coral-bed) is placed on the horizon of the English Coral Rag,
Coralline Oolite and Middle Calcareous Grit. In this district its
equivalents would therefore be the Corallian limestones of Upware
and the greater part of the Ampthill Clay. The Upper Coralline
Oolite (zone of *Pecten varians*) is considered to be of Upper
Calcareous Grit age; no representative of this occurs in the Cam-
bridgeshire district, unless the uppermost portion of the Ampthill
Clay be considered as such.

The Astartian is correlated with the same beds as those which
are similarly named in the Paris Basin, *i.e.* the Kimeridge Passage-
beds and the clays at the base of the Lower Kimeridge with
Rhynchonella inconstans and *Ostrea deltoidea.* It has already been
stated that the only representative of these beds in the district
under consideration is the *Deltoidea*-zone near Haddenham.

The Middle Kimeridge or *Pteroceras*-beds are considered to be the equivalents of the Lower Kimeridge of England and the *Virgula*-beds are correlated with the Upper Kimeridge. Dr Struckmann's correlation of the latter is scarcely applicable for the Cambridgeshire district. Out of the 20 species of fossils recorded from the so-called 'Upper Kimeridge' of Hanover, 9 occur in the Lower Kimeridge of England, whilst two only of these are found in the Upper Kimeridge, and one of them is more characteristic of the former than the latter subdivision.

The ' *Virgula*-beds' are therefore much more nearly allied to the Lower than to the Upper Kimeridge of England, and in my opinion should be placed in that division.

The Lower Portland beds of Hanover contain *Ammonites gigas*, and are correlated with the Portland beds. If this be their true position, there would be no beds in Hanover equivalent to our Upper Kimeridge.

In the following table I have endeavoured to shew the correlation of the Cambridgeshire Jurassics with their equivalents in the three continental areas with which they have been compared.

COMPARATIVE TABLE OF UPPER JURASSIC ROCKS.

CAMBRIDGESHIRE AND HUNTINGDONSHIRE.	UPWARE.	PARIS BASIN.	JURA.	HANOVER.
Discina-zone.		Lower Bolonian.	Portlandian.	Portlandian.
Virgula-zone.		Virgulian.	Virgulian.	Virgula-beds.
Alternans-zone.		Pteroceran.	Pteroceran.	Pteroceras-beds.
Astarte supracorallina-zone.		Astartian.	Astartian.	Astartian.
Deltoidea-zone.	Coral Rag.			
Phosphate Bed.	Coralline Oolite.	Corallian.	Corallian.	Coralline Oolite.
Ampthill Clay.	?	Oxford Oolite.		
	?	Oxford Grit.	Oxfordian.	Hersumer Beds.
St Ives and Elsworth Rocks.		Oxford Clay.	Le fer sous Oxfordien.	Ornatus-clays.
Oxford Clay.				

INDEX.

Kellaways Beds in Northamptonshire, 10
Kimeridge Clay, 61—76; basement-bed,
35, 61, 85; thickness, 74; list of fossils, 75; correlation, 85
Knapwell, 36, 41, 42, 61

Linton Hill, 82
List of works on the Jurassic Rocks of
Cambridge, 1
Littleport, 70—73
Longstanton, 36
Lower Calcareous Grit, 19—34; correla-
tion, 78

Market Rasen, 86

Needingworth, 36, 46
Neuvizy, 89

Oakington, 36, 61
Oakley, 82
Oolithe Corallienne, 91
Osmington, 82
Over, 36, 48
Oxford, 77
Oxford Clay, 9—18; thickness, 9; zones
in Northamptonshire, 11; fossils from
St Ives 17; zones in Cambridgeshire
and Huntingdonshire, 18; correlation,
77

Papworth St Agnes, 19
Papworth St Everard, 6, 41
Paris Basin, 87
Phillips, J., the basement bed of the
Kimeridge Clay near Oxford, 85
Phosphatic nodules, 23, 35, 36, 42, 47,
52, 61, 62, 63, 65, 68, 85
Pterocerian, 89

Rampton, 48
Rigaux, E., Elsworth Rock, 88
Roslyn (Roswell) Pit, 65, 74, 85

St Ives, 9, 15—18, 21, 44
St Ives Rock, 6, 21—25; list of fossils,
25
St Mihiel, 88
St Neots, 11, 13, 14
"St Neots Rock," 6, 13
Sandsfoot Castle, 82
Sandsfoot Grit, 83
Sandy, 14
Sedgwick, A., Corallian Rocks of Up-
ware, 57; Corallian between Hadden-
ham and Chatteris, 59; wells between
Cambridge and Lynn, 60
Seeley, H. G., the Fen-Clay, 5; Blun-
tisham Clay, 7; Tetworth Clay, 7;

Gamlingay Clay, 7; Ampthill Clay, 7;
Upware Limestone, 7, 57; classification of the Cambridgeshire Jurassic
Rocks, 7; St Neots Rock, 6, 13; Els-
worth Rock, 6, 19, 32; Bluntisham
cutting, 21; Holywell fossils, 23; the
relation of the Elsworth and St Ives
Rocks, 30, 34; Ampthill Clay at Els-
worth, 41; the rock at Boxworth, 44;
fossils from Bluntisham, 46
Sections: Clay-pit west of St Ives, 16;
St Ives to Bourn to illustrate Seeley's
views on the Elsworth and St Ives
Rocks, 30; St Ives to Elsworth, 31;
Upware ridge, 58
Selenite, 10, 14, 36, 39, 42, 43, 45, 47,
48, 64, 71
Septarian nodules, 15, 43, 45, 61, 64, 66,
70, 71
South Kelsey, 80
Spinney Abbey, 52
Stanton St John, 79
Stretham, 65
Struckmann, C., the Upper Jurassic of
Hanover, 91
Studley, 79, 81
Stuntney, 65
Subdivisions of the Jurassic Rocks in
Cambridgeshire, 8
Sub-Wealden boring, 79, 83, 86
Swavesey, 23, 31, 36, 45, 46

"Tetworth Clay," 7
Terrain à chailles siliceux, 91
Topley, W., the sub-Wealden boring, 83

Upware, 49, 51, 73, 84; fossils from the
south quarry, 53; fossils from the
north quarry, 56
"Upware Limestone," 7

Virgula-zone, 69, 85
Virgulian, 74, 89

Waddesdon Field, 82
Westcot, 82
Weymouth, 79, 82
Wheatley, 79
Wicken, 60
Wilburton, 65
Willingham, 36, 62
Worminghall, 82
Wrawby, 80
Wyke, 82

Zones:—in the Oxford Clay of East Northamptonshire, 11; in the Oxford Clay
of Cambridgeshire and Huntingdonshire, 18; in the Kimeridge Clay, 69

Printed in the United States
By Bookmasters